D0787426

Stationary Fuel Cells: An Overview

Stationary Fuel Cells: An Overview

Kerry-Ann Adamson

ELSEVIER

Amsterdam • Boston • Heidelberg • London • New York • Oxford
Paris • San Diego • San Francisco • Singapore • Sydney • Tokyo

Elsevier
Linacre House, Jordan Hill, Oxford OX2 8DP, UK
Radarweg 29, PO Box 211, 1000 AE Amsterdam, The Netherlands

First edition 2007

British Library Cataloguing in Publication Data
A catalogue record for this book is available from the British Library

Library of Congress Cataloging-in-Publication Data
A catalog record for this book is available from the Library of Congress

ISBN: 978-0-08-045118-3

Printed and bound in Great Britain
07 08 09 10 11 10 9 8 7 6 5 4 3 2 1

Contents

Acknowledgements

I would like to publicly thank all the companies and organisations contacted during the writing of this book that have provided images, hints or information.

Thank you to:

Ballard
BP
Ceramic Fuel Cells
Ceres Power
FuelCell Energy
H2PIA
Honda
Johnson Matthey Fuel Cells
Rolls Royce Fuel Cell
SFC SMART Fuel Cell
The International Energy Agency
The Smithsonian Institution
UK Department of Trade and Industry
US Department of Defense
UTC Fuel Cells

I would also like to thank Gemma Crawley, of Fuel Cell Today, for allowing me to use her work at will, Dr. Mike Hugh, Dr. Ben Todd and Dr. Jonathan Butler also of Fuel Cell Today for helping, editing and providing total support (and the odd PhD thesis to read!). I would also like to thank Dr. Peter Pearson, Imperial College, London, and Professor Georg Erdmann, TU Berlin, for still being mentors and mum, dad and my amazing sister Wendy.

List of figures

List of tables

Glossary

AFC	–	Alkaline Fuel Cell
AC	–	Alternating Current
APU	–	Auxiliary Power Unit
BOP	–	Balance of Plant
Btu/h	–	British Thermal Units per hour
CO	–	Carbon Monoxide
CO_2	–	Carbon Dioxide
CETI	–	Clean Energy Technologies (South Korean Company)
CHP	–	Combined Heat and Power
CDM	–	Clean Development Mechanism
CUTE	–	Clean Urban Transport for Europe
DBFC	–	Direct Borohydride Fuel Cell
DC	–	Direct Current
DFC	–	Direct Fuel Cell (NB: Term Specific to FuelCell Energy Product Line)
DFC/T	–	Direct Fuel Cell with Turbine (NB: Term Specific to FuelCell Energy Product Line)
DEFC	–	Direct Ethanol Fuel Cell
DG	–	Distributed Generation
DoD	–	Department of Defense (USA)
DoE	–	Department of Energy (USA)
DTI	–	Department of Trade and Industry (UK)
DMFC	–	Direct Methanol Fuel Cell
ECTOS	–	Ecological City Transport System
EU	–	European Union

FAFC – Formic Acid Fuel Cell
FCHU – Fuel Cell Heating Unit (NB Term Specific to Baxi innovation)
FCFI – Fuel Cell Financing Facility
FP7 – 7th Framework Programme (EU-Specific Term)
FCV – Fuel Cell Vehicle

Gt – Giga tonnes
GDL – Gas Diffusion Layer
GEF – Global Environment Facility
GPT – General Purpose Technology
GW – Giga watt

HFP – Hydrogen and Fuel Cell Technology Platform (EU-Specific Term)
HHV – Higher Heating Value
HTPEM – High-Temperature PEM

ICE – Internal Combustion Engine
IEA – International Energy Agency
IGFC – Integrated Gasification Fuel Cell (NB: Term Specific to Rolls Royce Fuel Cell)
IFC – International Finance Corporation
IP – Implementation Plan (of the Hydrogen and Fuel Cell Technology Platform (EU))
IPCC – International Panel Climate Change

JTI – Joint Technology Initiative (EU-Specific Term)

kWe – Kilowatt Electric
kWh – Kilowatt Per Hour
KEPCO – Korean Electric Power Corporation (South Korean Company)
KEPRI – Korean Electrical Power Research Institute (South Korean Company)
KIER – Korean Institute of Energy Research (South Korea)
KIST – Korean Institute of Science and Technology (South Korea)
KOGAS – Korean Gas Company (South Korean Company)

LCA – Life Cycle Analysis
LDV – Light-Duty Vehicle
LPG – Liquid Petroleum Gas
LHV – Lower Heating Value

mCHP	–	Micro Combined Heat and Power
MCFC	–	Molten Carbonate Fuel Cell
MEA	–	Membrane Electrode Assembly
MOST	–	Ministry of Science and Technology (South Korea)
MOCIE	–	Ministry of Commerce, Industry and Energy (South Korea)
MWe	–	Mega Watt Electric
MWth	–	Mega Watt Thermal
NEDO	–	New Energy and Industrial Development Organisation (Japan)
NEF	–	New Energy Foundation (Japan)
North America	–	USA and Canada
NPV	–	Net Present Value
NYPA	–	New York Power Authority
PAFC	–	Phosphoric Acid Fuel Cell
PEFC	–	Polymer Electrolyte Fuel Cell
PEM	–	Polymer Electrolyte Membrane (Fuel Cell)
POC	–	Proof-of-Concept
Pt	–	Platinum
R&D	–	Research and Development
RD&D	–	Research, Development and Demonstration
RECs	–	Renewable Energy Credits (Term Specific to the USA)
ROCs	–	Renewable Obligation Certificates (Term Specific to the UK)
RPG	–	Residential Power Generator
RPS	–	Renewable Portfolio Standard (Term Specific to the USA)
SECA	–	Solid Energy Conversion Alliance
SOFC	–	Solid Oxide Fuel Cell
TCO	–	Total Cost of Ownership
TFCPS	–	Toshiba Fuel Cell Power Systems
UPS	–	Uninterruptible Power Supply
VRLA	–	Valve Regulated Lead Acid (Batteries)

Preface

Since starting out in the fuel cell industry nearly a decade ago, I have seen the entire industry move from a nice idea to commercial proposition – in many applications at least. Whilst I, and many others, found my way here through an interest in fuel cell vehicles, it has been other applications, such as marine, residential and power plants, that have since taken hold of my imagination and day-to-day work life. Having read a substantial fraction of all the fuel cell, and hydrogen books, which have been published over the past 10 years, I came up with an idea to write a book, specifically tailored to potential adopters in the stationary sector, which was not a general introduction, of which there are many of varying qualities, and also not an engineering text. The aim was to write something useful that could be read during the commute home or in the evening and give enough information to allow people to make a first-cut decision as to whether it was worth their while looking further into fuel cells. Yes, each chapter could have been expanded twofold, threefold and upwards, and I am sure many people who read it will wonder why I did not include x, y or z, but the aim is to provide a food-for-thought text that lets people think 'can I ...'.

Kerry-Ann
Manager, Fuel Cell Today,
Informing the Fuel Cell Industry

1
Introduction

This book attempts to provide a range of potential adopters with unbiased and accurate information on the current state, and medium-term future, of fuel cell technology specifically for stationary applications. Many other applications for fuel cells do exist, such as fuel cell-powered light-duty vehicles, mobile phones, campervans and forklifts, but as the focus of this book is specifically building integrated and gird supporting units, these other applications are not covered.

The reason for this book is the increasing number of times Fuel Cell Today is being asked the question 'what do I need to know to help me decide if fuel cells are right in my case?'

If you are working as an architect, planner, plant manager or in a similar position, then this book attempts to answer the obvious questions on whether fuel cells might provide a suitable solution for your project's energy needs. The book itself is split into three main sections: (1) A beginner's guide to fuel cells; (2) Application chapters on uninterruptible power supply (UPS)/backup power, residential and non-residential/plant fuel cells; and (3) The political drivers behind the current drive towards cleaner, more efficient energy-producing technologies.

The time frame adopted by the book uses 2006 as a starting point and looks out to around 2015. This end date is used as many of the major driver nations behind the research and development (R&D) effort into fuel cells have detailed road maps projecting out to 2015. Where forecasts are included, wherever possible, the assumptions behind these are stated and the reference for each is clearly indicated.

What this book is *not* is a step-by-step installation guide. It will *not* give the self-builder the tools to install a fuel cell combined heat and power (CHP) unit, it will *not* provide wiring diagrams for a UTC PC25 200 kW phosphoric acid fuel cell, for example, and it will *not* provide a detailed examination of the differing routes of hydrogen production and how this could impact the overall economic and environmental cost of the units. For those interested in the highly technical engineering details behind each of the fuel cell types, there are already a number of weighty and informative texts on the marketplace, ranging in price, and coverage, from around £100 up to over a £1000. Where possible, a number of these texts have been indicated in the various chapters. These are not recommended texts, and there will be a number of other books out there that have not been listed. Also, due to the power of the internet, a substantial amount of further information can be found and readily accessed on the Web. Where documents have been used that may be of interest to for further reading the (current) location reference is given.

What needs to be kept in mind at all times though is that this is still a very young industry. Though the technology was invented in 1839 by William Grove, it has only really been since the 1990s that substantial R&D funding has been available to create the right conditions for technological breakthroughs. These breakthroughs have led to a number of fuel cell-based products being on, or close to, market ready (2006/2007), with others, such as fuel cell vehicles, appearing as far away as ever.[1] Because of the differing technologies[2] employed, the age of the industry and the comparatively low number of units installed, especially when compared with the main competing technologies, the learning curve[3] that the fuel cell industry is experiencing is still very steep. We are also in the challenging position in that it is not until these units enter large-scale testing or initial market penetration, which they are doing now for stationary applications, that many issues that we are facing can even be found to exist. In many ways, this means that the early adopters that are looking at using the technology now are the true trailblazers of this industry. Whilst this may excite some, it will also induce others, of the more cautious persuasion, to turn to more conventional tried-and-tested technologies.

Also as there are a number of types of fuel cells currently under development, some potential adopters will want to wait and see if there is one 'winner'

[1] Current (2006) projections from the automotive industry is somewhere between 2012 and 2015 for initial commercialisation and 2025 for full commercialisation.

[2] See Chapter 2 for an explanation of the differing technologies under development.

[3] A learning curve is the rate of learning experienced by a new industry. A good entry book into this fascinating area is the OECD/IEA book "*Experience Curves for Energy Technology Policy*", which is available online at http://195.200.115.136/textbase/nppdf/free/2000/curve2000.pdf.

technology (i.e., a type of fuel cell that becomes locked-in[4]). At present, there are no indications of this lock-in effect happening. In terms of sheer numbers, the low-temperature fuel cells, proton-exchange membrane fuel cells (PEMFCs) and direct methanol fuel cells (DMFCs), are currently dominating, but MCFC, PAFC and SOFC have a number of advantages that ensure that they will continue to be used and developed as the markets grow and diversify.

1.1 Market development to date

This section of the book sets the scene of current activity in this specific area and puts it in the context of the wider fuel cell industry. Figure 1.1[5] shows fuel cell market development to date, with a breakdown of stationary and non-stationary applications, whilst Figure 1.2 breaks the stationary sector down further into regional activity.

What these figures show is that not only are stationary developments increasing, but their overall market share as the industry expands is also increasing. The stationary sector is now sitting at more than 20% of all installed fuel cell units to date. Geographically, Japan by far and away dominates, in terms of units installed. As will be discussed at a number of points in this book,

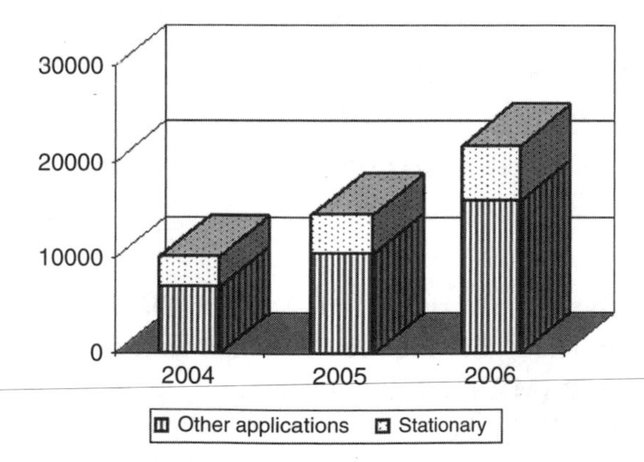

Figure 1.1 Fuel cell market to date by numbers
Source: © Fuel Cell Today

[4] Lock-in exists when one technology is so pervasive in the marketplace that no other technology can enter or compete against it. The other technologies are effectively locked-out of the market.

[5] All the figures in this section are taken from Fuel Cell Today annually updated application surveys that are available online at http://www.fuelcelltoday.com.

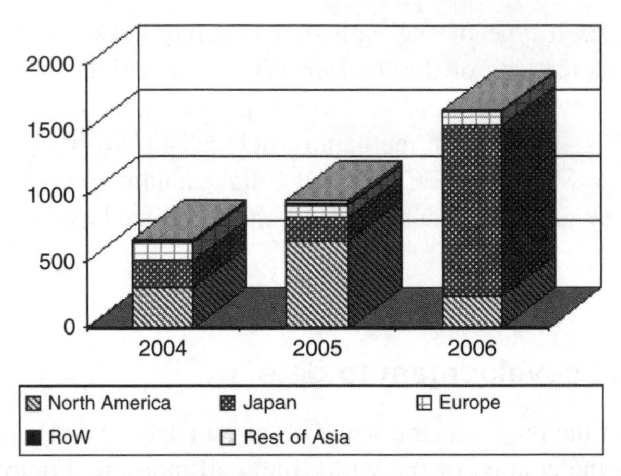

Figure 1.2 Stationary fuel cell market to date by region and number of units installed
Source: © Fuel Cell Today

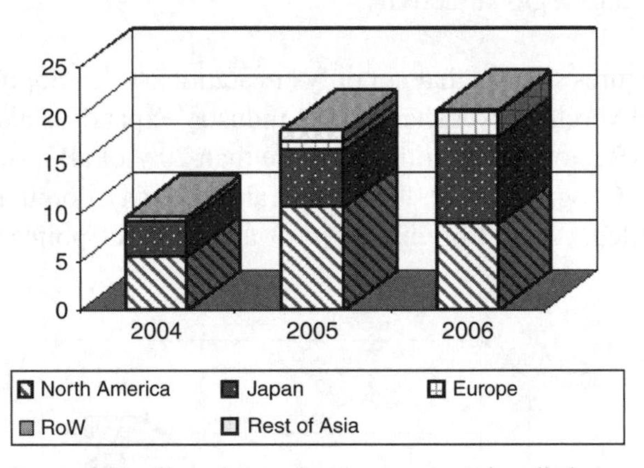

Figure 1.3 Stationary fuel cell market to date by megawatt installed
Source: © Fuel Cell Today

this is due largely to the very proactive stance that the Japanese government has taken in residential fuel cell development. If we switch the view though to megawatt (MW) installed capacity by region, Figure 1.3 shows that North America (NA), here classed as the USA and Canada, and Japan are on an equal footing, indicating that to date NA is currently more open to the adoption of larger units in terms of kilowatt electric (kWe) output.

In short, what this shows is that the stationary applications are increasingly buoyant, and we expect this to continue. This has been achieved not just through government support but also because the technology is able to provide a win-win situation for adopters, especially in the growing UPS sector.

1.2 Summary

Most commentators in the field agree that fuel cells have a place in the future energy landscape as a highly efficient power generation technology. They are not a proverbial 'silver bullet'[6] and should not be thought of as such. They will have a place in a range of technologies, using a range of fuels, but they will not provide a single answer to the energy problems currently faced by nations across the world.

What this all adds up to is no simple, single answer. As the markets grow and the degree of product standardisation in the fuel cell industry increases, adoption will become more routine, but it is always important to look at all the options, including the current incumbent technologies, available, and this means not just fuel cells.

To those trailblazers who do not want to wait until the standardised mass market has arrived, then this book should arm you with enough information to ask the right questions to see if fuel cell technology is right in your case.

[6] A silver bullet is a metaphor used to describe a perceived simple single solution to a complex problem.

2
Fuel cell basics

2.1 Fuel cell technology

Although this book is about end-use adoption, it is useful to have a basic grounding in the technology in question. This chapter aims to provide a run-through of fuel cell technology and its main application areas. If by the end of the chapter it sparks an interest to learn more, there is a run down of a number of popular engineering-based text books.

2.2 What is a fuel cell?

A fuel cell is an electrochemical conversion device that relies upon a continuous feed of a fuel, ultimately hydrogen, to produce DC electricity and, as by-products, heat and water.

Fuel cells work on a basic chemical principle and use some very modern engineering. A cell contains an anode, a cathode and an electrolyte layer, with a typical system being shown in Figure 2.1.

The electrolyte allows the passage of the positively charged hydrogen ions, whilst the negatively charged electrons pass around an external circuit, thus creating an electric current. At the interface with the cathode, the catalyst creates a reaction with oxygen during which water and (due to the process being exothermic) heat are produced.

As each individual cell only produces a very low wattage, around 1 V, a fuel cell stack is made up of a number of these base units linked in series (Figure 2.2).

A fuel cell system comprises the stack, balance-of-plant (BOP), a DC–AC converter and, if required, a fuel reformer all packaged together. The fuel

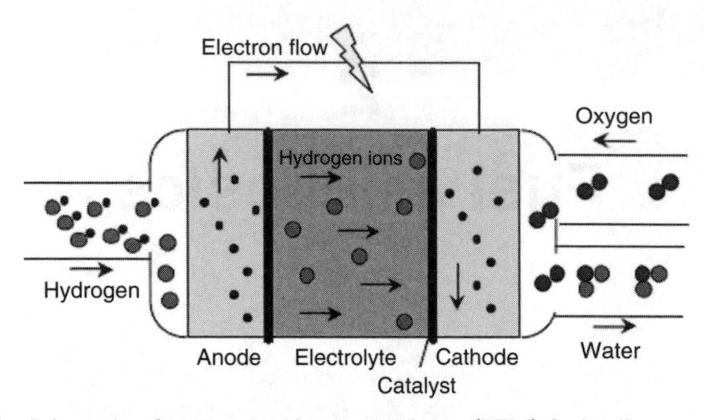

Figure 2.1 Schematic of a proton-exchange membrane (PEM) fuel cell
Source: © Fuel Cell Today

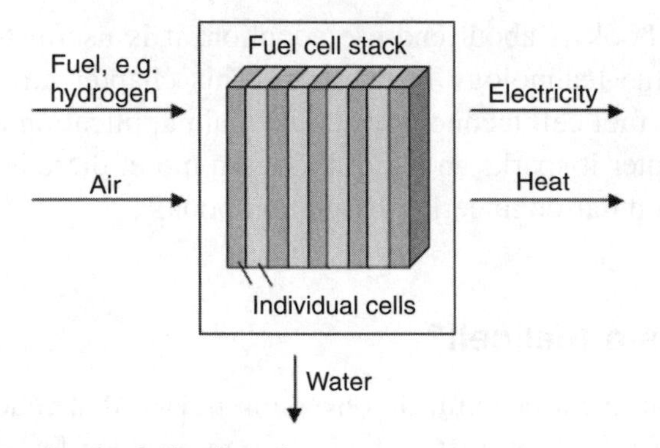

Figure 2.2 Schematic of a fuel cell stack
Source: © Fuel Cell Today

cell systems are then integrated into a wide variety of products. One such is shown in Figure 2.3.

What a fuel cell is *not* is a battery or an internal combustion engine. With a fuel cell as long as there is a fuel supply it will keep producing electricity, whereas a battery only has a defined unit of fuel, which once exhausted stops producing electricity. In a fuel cell, the conversion to power is through electrochemical means, whilst in an engine it is by combustion.

2.3 Fuel cell components

Going back down to the 1V base unit of the fuel cell, this itself is made up of a number of layers, which are listed below.

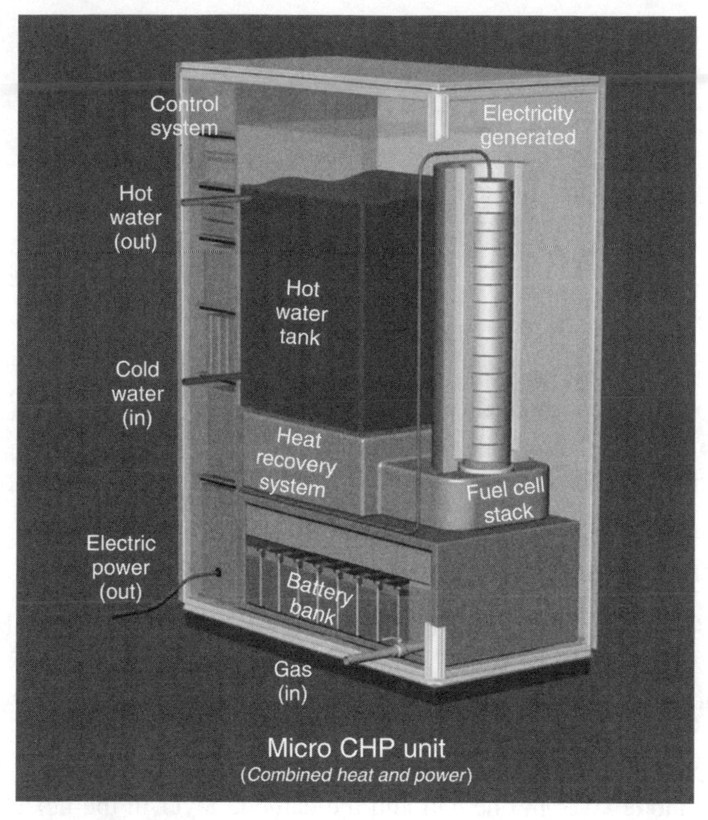

Figure 2.3 Electricity generator fuel cell system
Source: © Ceramic Fuel Cells

2.3.1 Electrolyte

The electrolyte is a material that allows the passage of the hydrogen ions, whilst blocking the passage of the negatively charged electrons. The specific type of material depends on the type of fuel cell. The main types are the flexible membrane (an example is shown in Figure 2.4) or the solid-state membrane.

2.3.2 Catalyst

Some high-temperature fuel cells do not need a catalyst to kick-start the reaction, but the ones that do use a range of chemical agents, the most common of which is the precious metal platinum (Pt).

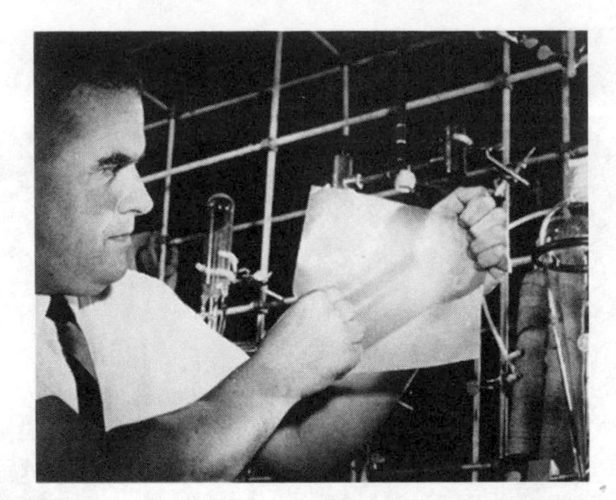

Figure 2.4 General Electric electrolyte in the 1960s
Source: © Smithsonian Institution

2.3.3 Membrane electrode assembly (MEA)

The MEA is a 'sandwich' made of the electrolyte membrane, located between the gas diffusion layer (GDL) and catalyst, with two field flow plates on the outside. In Figure 2.5, the input fuel flows from the left.

Bipolar plates are the end plates (shown above as field flow plates) which act as an anode layer in one cell and the cathode layer in the next.

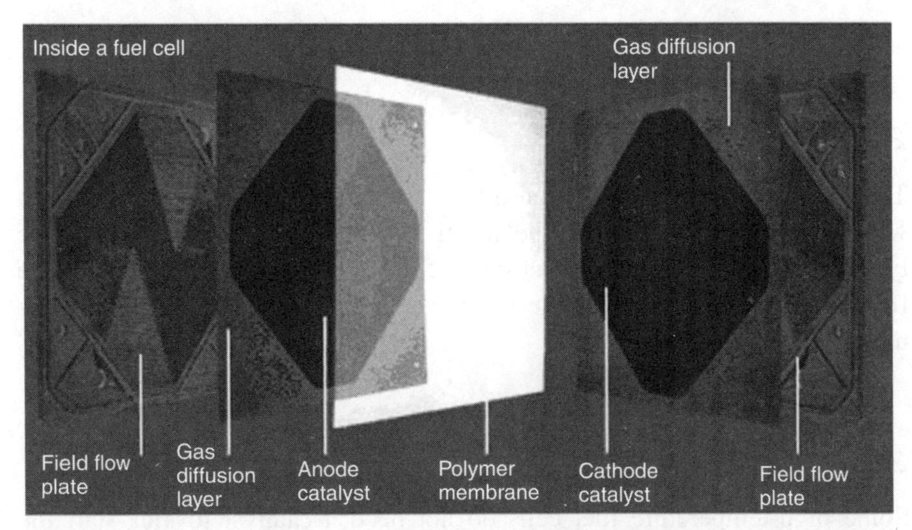

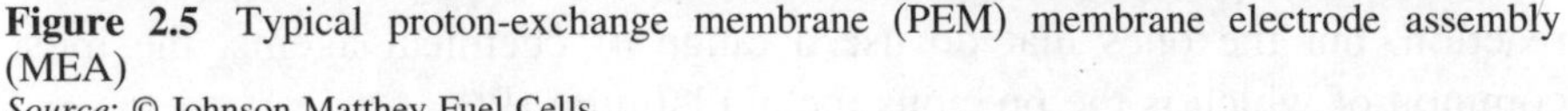

Figure 2.5 Typical proton-exchange membrane (PEM) membrane electrode assembly (MEA)
Source: © Johnson Matthey Fuel Cells

Apart from these base components, the fuel cell also contains current collectors, heat exchangers and manifolds.[1]

2.4 Types of fuel cells

As has already been mentioned, there are a number of different types of fuel cells. In alphabetical order, the most common are, alkaline fuel cells (AFCs), direct methanol fuel cells (DMFCs), molten carbonate fuel cells (MCFCs), phosphoric acid fuel cells (PAFCs), proton-exchange membrane fuel cells (PEMFCs), and solid oxide fuel cells (SOFCs). Apart from DMFC, which is technically a subset of the polymer electrolyte membrane fuel cells, others are named after their electrolyte type.

Given below is a brief overview of each of the different fuel cell types.

2.4.1 AFC

AFCs have been around the longest of any of type of fuel cell. Developed in the 1930s by Francis Bacon, it was this type of fuel cell that was used to power a number of the very early fuel cell vehicles (FCVs). It was also this type of fuel cell that caused the infamous "Houston we have a problem" statement made by Captain of Apollo 13, James Lovell.

Technological overview (Gemma Crawley, Fuel Cell Today)

The alkaline fuel cell (AFC) uses an alkaline electrolyte such as potassium hydroxide (usually in a solution of water) in order to operate. AFC systems are classified as low-temperature fuel cells and usually operate between 60°C and 90°C. AFCs use a variety of metals to speed up the reactions at the anode and cathode, with Nickel being the most commonly used catalyst.

Due to the rate at which the chemical reactions take place within the cell, AFC systems usually demonstrate efficiencies between 45% and

[1] Current collector is an inert structure of high electrical conductivity that is used to conduct current to or from an electrode. Heat exchangers are devices that are used to transfer heat from one material, or phase, to another. Manifolds are flat structures that have been bent. The function of the manifold is to direct, primarily, fluids around the cell.

60%. AFCs can produce up to 20 kW of electric power, and some newer designs have been reported to operate at temperatures as low as 23–70°C.

The disadvantage of using AFCs is that the strongly alkaline electrolytes adsorb even the smallest amount of CO_2 which, in turn, eventually reduces the conductivity of the electrolyte. (A number of AFC companies have claimed that this effect is not irreversible as the electrolyte can be changed.) For effective operation, it is also necessary to purify the oxygen used in the cell, and together these purifications can be very costly.

Also as the electrolyte material is corrosive, and being in liquid form, it makes the sealing of the anode and cathode gases more problematic than when a solid electrolyte is used.

Though this type of fuel cell is well known and understood, it currently (up to 2007), in terms of numbers, makes up less than 5% of new units produced each year, shown in Figure 2.6.

Current manufacturers in this area include Cenergie and HydroCell.

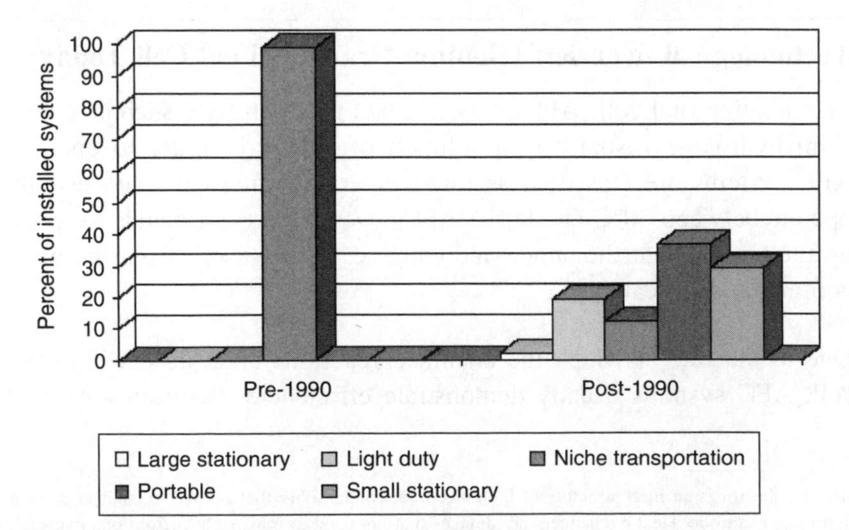

Figure 2.6 Historical alkaline fuel cell (AFC) market development
Source: © Fuel Cell Today

2.4.2 DMFC

DMFCs are the only mainstream form of low-temperature fuel cells, to date, to accept a fuel directly that is not hydrogen. They are commonly classified as a subcategory of PEMFCs, as they use the same internal structure and makeup, though in any thorough form of analysis they are separated out as a stand-alone type.

The main difference between a PEM system and a DMFC unit is that the waste products from the reaction of DMFC includes carbon dioxide.

Technological overview (Gemma Crawley, Fuel Cell Today)

Direct methanol fuel cells (DMFCs) employ a polymer membrane as an electrolyte. The system is a variant of the polymer electrolyte membrane (PEM) cell; however, the catalyst on the DMFC anode draws hydrogen from liquid methanol. This action eliminates the need for a fuel reformer and allows pure methanol to be used as a fuel.

The pure methanol is mixed with steam and fed directly into the cell at the anode. Here, the methanol is converted to carbon dioxide and hydrogen ions. The electrons are then pushed round an external circuit to produce electricity (before returning to the cathode), whilst the hydrogen protons pass across the electrolyte to the cathode, as occurs in a standard PEM fuel cell. At the cathode, the protons and electrons combine with oxygen to produce water.

The operating temperature for DMFCs is in the range of 60–130°C but is typically around 120°C, producing an efficiency of about 40%. Due to the low temperature conversion of methanol to hydrogen and carbon dioxide, the DMFC system requires a noble metal catalyst.

In the mid-1990s, DMFC units were very much in vogue among developers as a potential fuel cell for light-duty vehicle propulsion. This concept went out of fashion a few years later, but research and development (R&D) work continued apace and they are now seen as one of the main technologies,

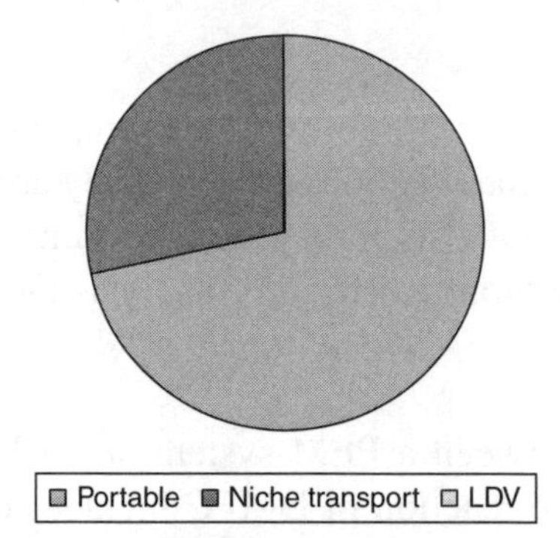

Figure 2.7 1996–2006 direct methanol fuel cell (DMFC) development by application
Source: © Fuel Cell Today

along with PEMs, for personal electronic fuel cells such as laptops and mobiles. There is also work in this area on using DMFCs as auxiliary power units (APUs), with SFC SMART Fuel Cell (Germany) being one of the world's frontrunners in this area.

In terms of DMFC market growth, Figure 2.7 shows the last ten years' development history. What is overridingly clear is that just two applications dominate. Portable, primarily personal electronics and niche transport (defined here as the use of a fuel cell, such as APU use, in any vehicle that is not a car or bus) have both experienced significant adoption of DMFCs.

Current developers of DMFC include SFC SMART Fuel Cell, Motorola Labs, MTI Micro, Ultracell and Toshiba.

2.4.3 MCFC

MCFC is one of the key technologies for stationary applications and is currently being deployed globally. Current development work is based on a modular 250 kW size base unit; both hybridised with a gas turbine for quick start-up, as well as solely as a self-contained fuel cell power system. One of the main benefits of MCFCs is that they can handle hydrogen from a range of fossil fuel-based sources without the need for extraneous fuel-conversion devices. Its operating temperature is the highest when compared to that of fuel cells under development, with current R&D efforts focusing on bringing costs down, increasing the efficiency of fuel conversion, increasing the durability of sealants and testing multi-megawatt plants (Figure 2.8).

Figure 2.8 Artist's schematic of a fuel cell energy molten carbonate fuel cell (MCFC) power plant
Source: © Fuel Cell Energy

Technological overview (Gemma Crawley, Fuel Cell Today)

Molten carbonate fuel cells (MCFCs) are high-temperature systems that use an immobilised liquid molten carbonate salt as the electrolyte. Salts commonly used include lithium carbonate, potassium carbonate and sodium carbonate. Typically MCFC units have an operating temperature of around 650°C and an efficiency of around 60%. (This can rise to as much as 80% if the waste heat is used for cogeneration.)

Upon heating, the salts melt and generate carbonate (carbon trioxide) ions. These ions flow away from the cathode and towards the anode where they combine with hydrogen. This produces water, carbon dioxide and electrons. The electrons are passed through an external circuit, generating electricity and eventually returning to the cathode. At the cathode, oxygen and carbon dioxide (which has been recycled from the anode) react with the electrons to form carbonate ions that replenish the electrolyte.

There are several advantages associated with the MCFC system. First, the high operating temperature dramatically improves reaction kinetics and removes the need for a noble metal catalyst. The higher temperature also makes the cell less prone to carbon monoxide poisoning than lower temperature systems. As a result, MCFC systems can operate on a variety of different fuels.

Disadvantages associated with MCFC units arise from using a liquid electrolyte rather than a solid and the requirement to inject carbon dioxide at the cathode as carbonate ions are consumed in reactions occurring at the anode. There have also been some issues with high-temperature corrosion, but this can now be controlled to achieve a practical lifetime.

In terms of development numbers, Figure 2.9 serves to reiterate the fact that MCFCs are being developed for large stationary applications. Note that in Figure 2.9 anything above 10 kW is classed as a large stationary unit.

Ansaldo, FuelCell Energy and MTU CFC Solutions, which uses fuel cell stacks from FuelCell Energy, are the main players in this area.

2.4.4 SOFC

If this book had been published only five years ago, then SOFC would have been in the same category as MCFC units – i.e., large stationary only. Since then, there has been a substantial, and successful, amount of work done on decreasing the operating temperature and size of SOFCs with the impact they are moving away from solely large stationary units into other areas, shown later graphically (Figure 2.10).

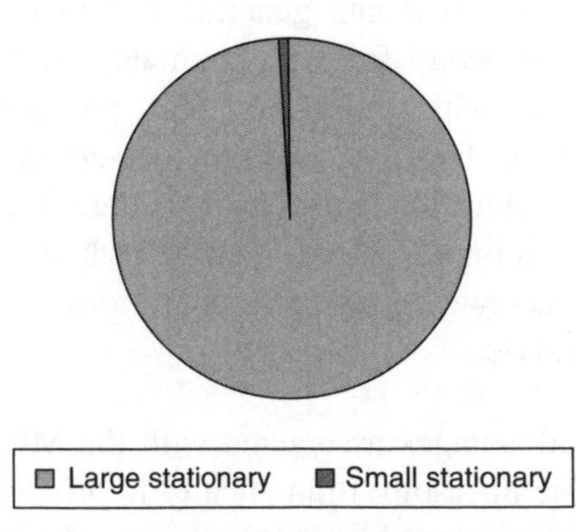

Figure 2.9 1996–2006 molten carbonate fuel cell (MCFC) development by application
Source: © Fuel Cell Today

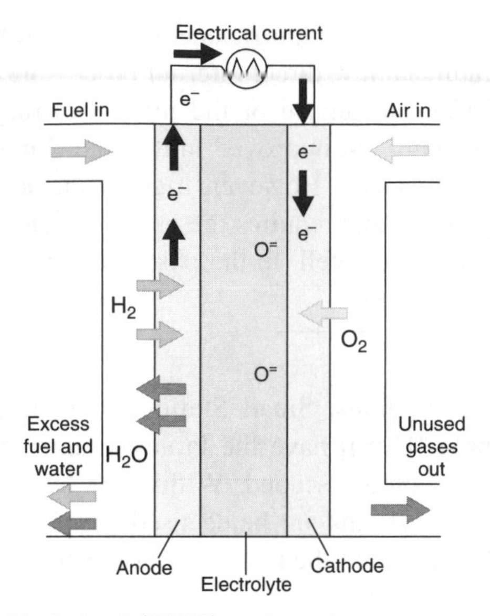

Figure 2.10 Solid oxide fuel cell (SOFC) – schematic
Source: © United States Department of Energy

Technological overview (Gemma Crawley, Fuel Cell Today)

Solid oxide fuel cells (SOFC) are high-temperature fuel cells that operate in the range of 800–1000°C. The SOFC is based around the same central design as many other fuel cell technologies and is composed of an anode and a cathode separated by an electrolyte. In the case of the SOFC unit, this electrolyte is a solid ceramic, such as zirconium oxide stabilised with yttrium oxide. During operation of the cell, oxygen (in the form of air) is supplied at the cathode. The ceramic electrolyte conducts oxygen ions from the cathode to the anode, whilst electrons are pushed round an external circuit in order to produce electricity. At the anode, the oxygen ions combine with hydrogen to produce water. Heat and carbon dioxide are also generated at the anode.

There are two basic designs of SOFC units. In the planar design, components are assembled in flat stacks where the air and hydrogen traditionally flow though the unit via channels built into the anode and cathode. In the tubular design, air is supplied to the inside of an extended solid oxide tube (which is sealed at one end), whilst fuel flows round the outside of the tube. The tube itself forms the cathode, and the cell components are constructed in layers around the tube.

SOFC units have achieved high efficiencies of over 60%, and the high operating temperatures allow direct internal processing of fuels such as natural gas. A further advantage of the high operating temperature is that the reaction kinetics are improved in the SOFC unit thus removing the need for a metal catalyst. However, high-temperature corrosion can sometimes be a problem and requires the use of expensive materials and protective layers within the cell. In the past, there have also been some problems related to sealing.

In terms of numbers of units, Small Stationary [residential and uninterruptible power supply (UPS)] have the largest number of units deployed, with Niche Transport coming second. Within the Niche Transport sector, the majority of SOFC systems are being used as APUs (Figure 2.11). As residential and UPS are two of the main foci of the book, we shall deal with them in more depth later on.

A number of companies are working with SOFCs. Some of the main players relevant to this text are Ceramic Fuel Cells, Ceres Power, Kyocera, MHI, Rolls Royce Fuel Cells, Siemens Westinghouse and Versa Power Systems (Figures 2.12 and 2.13).

2.4.5 PAFC

Like AFCs, PAFCs have been around for a comparatively long time. Also like AFCs, interest in PAFCs seemed to wane in the late 1990s, but now they are starting to experience something of a renaissance. Working within

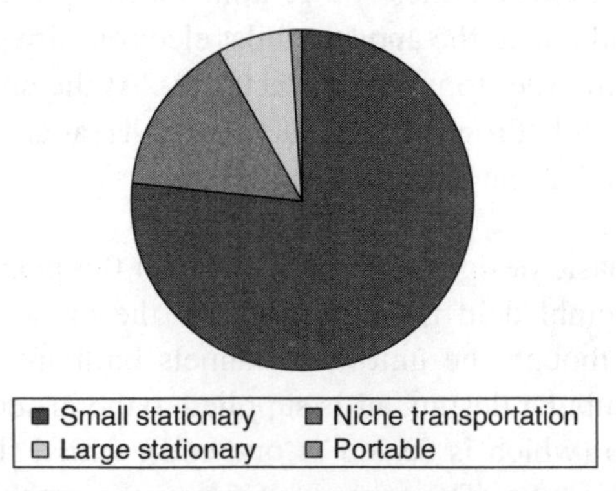

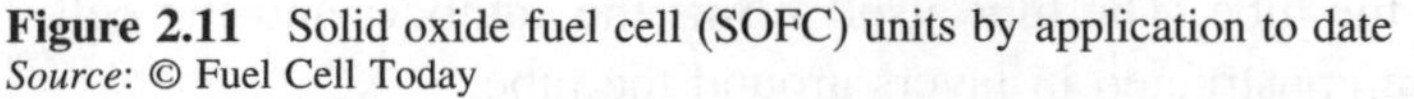

Figure 2.11 Solid oxide fuel cell (SOFC) units by application to date
Source: © Fuel Cell Today

Figure 2.12 Ceramic Fuel Cells solid oxide fuel cell (SOFC) building block – schematic
Source: © Ceramic Fuel Cells

Figure 2.13 Ceramic Fuel Cells solid oxide fuel cell (SOFC) building blocks
Source: © Ceramic Fuel Cells

the same space as MCFCs, PAFCs are ideally suited to large stationary applications where they can produce electricity as well as high-grade heat (Figure 2.14).

Technological overview (Gemma Crawley, Fuel Cell Today)

A phosphoric acid fuel cell (PAFC) consists of an anode and a cathode made of finely dispersed platinum catalyst on a carbon and silicon carbide structure that holds the liquid phosphoric acid electrolyte. Typically, PAFC systems have an operating temperature of around 200°C. When used for the co-generation of electricity and heat, PAFC cells can be around 85% efficient. However, when used for the generation of electricity alone, this efficiency drops to between 35% and 40%.

Within the system, positively charged hydrogen ions pass through the phosphoric acid electrolyte, moving away from the anode and towards the cathode. Electrons generated at the anode are passed through an external circuit, producing electricity and returning to the cathode. At the cathode, the electrons and hydrogen ions combine with oxygen to form water.

Whilst the efficiency of PAFCs is lower than that of other fuel cell types, there are a number of advantages associated with the systems. For example, PAFC systems are simple to construct, stable and exhibit low electrolyte volatility. In addition, by operating at a temperature of around 200°C, PAFC units can tolerate a carbon monoxide (CO) concentration of about 1.5%, making them less susceptible to CO poisoning.

Disadvantages associated with PAFC models include the necessity to remove sulphur from the fuel in order to avoid damaging the electrode catalyst. In addition, the use of an acid for the electrolyte requires the other components in the cell to be corrosion resistant.

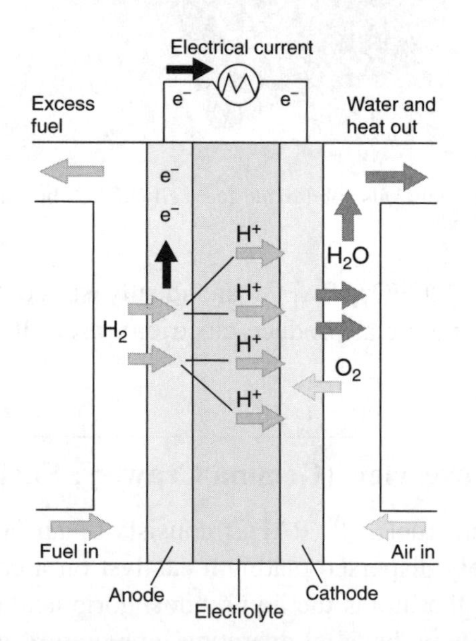

Figure 2.14 Phosphoric acid fuel cell (PAFC) schematic
Source: © United States Department of Energy

Figure 2.15 Honda FCX fuel cell vehicle (Powered by a PEMFC)
Source: © Honda

In terms of growth, since 1996 the market has seen over 250 separate units installed, and all, with just one exception, in the large stationary sector.

The predominant producer to date of PAFC units is UTC Power, daughter company of the international goliath United Technologies Corporation of the USA. Dozens of these UTC units have been up and running globally for a substantial period of time, with some media-friendly success stories. The incident most frequently quoted comes from the New York Police Department station in Central Park, New York. The station has a UTC PureCell 200 (formerly known as a PC25) 200 kW unit installed. During the infamous 2003 blackout, which knocked power out across most of Manhattan, the PureCell unit kept running smoothly, producing power. Anecdotal evidence suggests that the police on duty only realized that there was a blackout when they looked out the windows to see the city in darkness!

2.4.6 PEMFC

PEMFCs are by far and away the best known of the fuel cell stack types. Though focused research on PEM fuel cells only started intensively in the 1990s at Ballard (Canada), the technology has experienced large leaps forward. Within this, the best known of the PEM-based applications is light-duty FCVs, but there are also increasing numbers of units employed in such diverse areas as forklifts, residential units, laptops, yachts and even robots. (Figure 2.15)

Technological overview (Gemma Crawley, Fuel Cell Today)

The proton-exchange membrane (PEM) fuel cell uses a thin, permeable polymeric membrane as the electrolyte. The membrane is very small and light and in order to catalyse the reaction, platinum electrodes are used on either side of the membrane. Within the PEM fuel cell unit, hydrogen molecules are supplied at the anode and split into hydrogen protons and electrons. The protons pass across the polymeric membrane to the cathode, while the electrons are pushed round an external circuit in order to produce electricity. Oxygen (in the form of air) is supplied to the cathode and combines with the hydrogen ions to produce water.

Compared to other electrolytes (which require temperatures up to 1000°C to operate effectively), PEMFCs operate at very low temperatures of about 80°C allowing rapid start-up. The efficiency of a PEM unit usually reaches between 40% to 60% and the output of the system can be varied to meet shifting demand patterns. Typical electric power is up to 250 kW. In addition, PEM fuel cells are often compact and lightweight units.

As the electrolyte is a solid rather than a liquid, the sealing of the anode and cathode gases is far easier and this in turn makes the unit cheaper to manufacture than some other types of fuel cells. Furthermore, the solid electrolyte can lead to a longer cell and stack life as it is less prone to corrosion than some other electrolyte materials. However, there are also some disadvantages associated with PEM operation.

Although the low operating temperature of the unit is usually seen as an advantage, in some instances temperatures as low as 80°C are not high enough to perform useful cogeneration. Furthermore, in order to achieve the most effective operation of the unit, the electrolyte must be saturated with water. Control of the moisture of the anode and cathode streams therefore becomes an important consideration. The PEM fuel cell is also sometimes referred to as a polymer electrolyte fuel cell (PEFC).

In terms of numbers, PEM units have outstripped all other types of technologies to date. Figure 2.16 shows that the vast majority of these are in the portable sector, here defined as anything designed to be picked up and moved by hand, with stationary units making up just over a quarter.

Though these are the main types of fuel cells, there are a number of other varieties, some exotic and some currently R&D curiosities that might one day become mainstream. These fringe forms of fuel cell include biological fuel cells, direct borohydride fuel cells (DBFCs) (technically a subcategory of AFCs), direct ethanol fuel cells (DEFCs) (a further subcategory of PEM fuel cells), and formic acid fuel cells (FAFCs). One other type of unit that is sometimes classed as a fuel cell are metal-air units, which use solid metal bars, usually a form of aluminum, as the fuel instead of hydrogen.

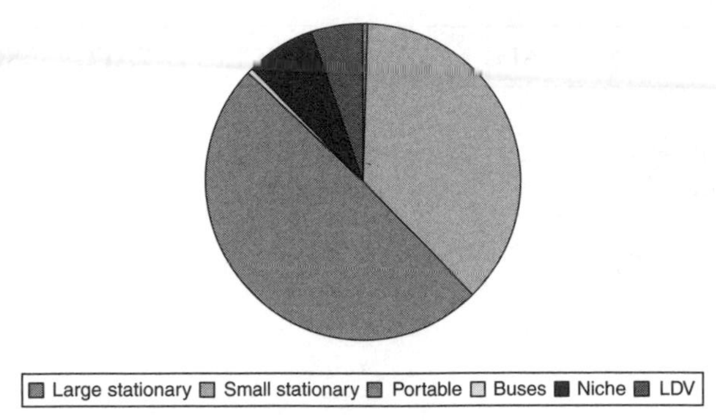

Figure 2.16 1996–2006 development of proton-exchange membrane (PEM) units
Source: © Fuel Cell Today

2.5 Fuel cell applications

Name a product that needs electricity as an input and you can almost certainly guarantee that someone somewhere is looking, or has looked, at using a fuel cell to power it. These can be anything from the comparatively mundane, such as providing power to the home, to the highly imaginative, such as providing power to a proposed NASA Lunar Base. As the fuel cell market is still far from fully developed, and many applications are still in the prototype phase, the full range of these potential application areas has yet to be understood.

Most analysts tend to classify the types of applications in a small number of pockets: portable, which covers personal electronics and units that have been designed to be moved around; UPS systems; residential fuel cells; power plant fuel cells; light-duty FCV; fuel cell buses and niche transport, which covers everything that is not a LDV or bus and APUs.

Though there are six main types of fuel cells, each has a small number of applications that it is best suited to. The matrix below shows the current perceived wisdom on which fuel cell fits best into which application area.

What can immediately be seen from this table is that PEM fuel cells appear to have the potential, at some point in the future, to be something of a general purpose technology (GPT), whilst technologies such as MCFC and PAFC will remain important in large stationary applications, but it is highly unlikely that we will see them in light-duty vehicles in the future.

The rest of this chapter outlines three applications, namely APUs, buses and light-duty vehicles that are of general interest but not covered in this book.

	AFC	DMFC	MCFC	PAFC	SOFC	PEM
Transport						
Large transport, such train or marine	X		X	X	X	X
Bus						X
Small-utility vehicles	X	X				X
Light-duty vehicles						X
Auxiliary power unit	X	X			X	X
UPS						X
Residential					X	X
Non-residential buildings			X	X	X	X
Power plant			X	X	X	X
Electronics		X				X
Portable		X			X	X

2.5.1 APUs

An APU is a unit on board a vehicle that is used to provide non-propulsive power. The term initially came out of the aerospace industry but has now filtered into common usage for technology that provides the hotelling,[2] in the form of electricity, or heat and power needs, of a vehicle. When you consider the number of types of vehicles that require some form of on-board electrical load, you then start to realise that the market potential for an efficient electricity, and sometimes heat, generator is significant. Planes, trucks, cruise ships, yachts, cars, campervans, armoured vehicles, coaches and the list goes on. At present, this hotelling is provided by three types of technologies. The first is basic parasitic load that derives from the propulsion unit, second is an on-board battery, with all the inbuilt issues of using battery technology, and the third is by using an on-board generator.

Fuel cells are in no way being promoted as the only technology for all APU applications, with many issues still to be overcome, but with around a dozen applications, including all of those listed in the first paragraph, being actively worked upon to date, this will become a substantial market area for the new technology (Figure 2.17).

The military is a big player in this application, as it is with many others in the fuel cell space. One example of a project coming under military

[2] Hotelling is a term referred to the on-board electrical power load required by the vehicle for all systems other than propulsion.

Figure 2.17 SFC SMART fuel cell auxiliary power unit (APU)
Source: © SFC Smart Fuel Cell AG

guise is 'Silent Watch'. Hydrogenics, a USA-based fuel cell manufacturer, has been contracted to manufacture a regenerative PEM fuel cell for light armoured vehicles (LAV), specifically the Stryker unit. The 'multi-service-regenerative fuel cell auxiliary power unit' (MREF-APU), as it is being called, is being developed to meet low emission (acoustic, thermal, pollutants, etc.) requirements for reconnaissance missions for the Stryker LAV. The APU produces hydrogen through the electrolyser, and then when the engine is switched off, the PEM fuel cell uses the hydrogen, stored in a metal hydride, for powering electrical equipment. This neatly packaged unit will help to provide the army with increased stealth capabilities.

2.5.2 Buses

Buses are a socially good thing. They are something politicians as well as taxpayers find easy to support. When this is coupled with fuel cell technology and its potential to provide emission-free, vibration-free, silent buses, then this is a win-win. To date, there has been a number of very high-profile fuel cell bus trails, the largest by far being the Clean Urban Transport for Europe (CUTE) programme. CUTE began in November 2001 and continued until May 2006, during which time 27 fuel cell buses were operated in nine participating cities across Europe. In addition, a further six buses run within the associated programmes Ecological City Transport System (ECTOS), based in Reykjavik, Iceland also funded by the EU, and the STEP programme in Perth, Western Australia.

The aim of the CUTE project was to develop and demonstrate an emission-free and low-noise transport system that would contribute to cleaner environmental conditions, increase public knowledge and acceptance of hydrogen and fuel cell technology and build a strong foundation for regulation and certification of fuel cell technology. It was widely agreed that throughout the course of the programme, the fuel cell buses had performed beyond expectations and the project had been successful in demonstrating fuel cell technology as a solution to mass public transport requirements.

The success of the CUTE, ECTOS, and STEP bus demonstration projects led to a one-year extension for the schemes. Cities participating in the contract extension included Amsterdam, Barcelona, London, Luxemburg, Madrid and Reykjavik, and each of these locations continued to operate three fuel cell buses throughout 2006. Hamburg also participated in the extension, increasing its fleet size from three to nine and in doing so, making it the world's largest fuel cell bus fleet being operated by a single transit agency. Continued operation of the buses into early 2007 and the ongoing operation of six more Mercedes-Benz Citaro buses in Perth and Beijing and three Gillig (a manufacturer) buses running in California is expected to continue the push towards commercialisation in this application.

2.5.3 Light-duty vehicles

Light-duty FCV has been the application that has experienced the most hype, most column inches and most nonsense written about it. To date, there has been in the region around 600 vehicles produced globally, with 2015 now being seen as a realistic date for some form of initial commercialisation. Within the vehicle, the fuel cell provides electric power for traction, thereby defining the vehicle as an electric vehicle, and often all the onboard electrical power requirements as well. The fuel, hydrogen, is stored in either compressed gas tanks or as a liquid fuel (Figure 2.18).

Cars are seen as one of the most challenging applications for fuel cells (along with aerospace) and so far still face a number of technical, infrastructural and regulatory barriers before we can expect to see full-scale commercialisation. Looking at the manufacturers, all the major global automotive players are involved in fuel cell development. Among them, only BMW is looking at using fuel cells as on-board APUs, whilst the others are developing fuel cell technology for primary propulsion.

In terms of competition for fuel cells in light-duty vehicles, it is easy to ignore this area and assume that the one you are backing is by-far-and-away

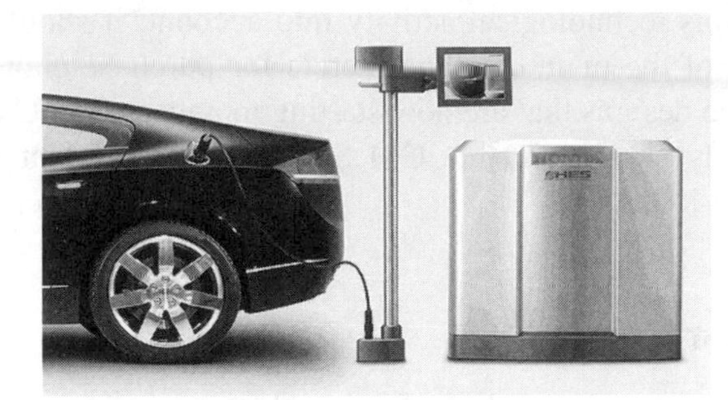

Figure 2.18 Honda home refuelling station
Source: © Honda

the best, and obvious, choice. For light-duty vehicles, not only is the current technological suite still continuously improving but a growing number of car manufacturers are looking at other options also, in parallel to fuel cell development. The two main other option here are the continued adoption of hybrid technologies and the development of the hydrogen-fuelled internal combustion engine.

Hybrids, depending on who you listen to, are rapidly becoming the consumers' favourite, or are fast falling out of favour with the very same consumer. What we do know is that sales of hybrids are continuing to grow, and now all major automotive manufacturers have some form of hybrid development programme. What seems to be becoming clear though is that hybrids are not, as first thought, a direct competitor to fuel cells but instead are a technological stopgap until FCVs reach mass-market readiness. They may, therefore, in fact be buying time for companies to further develop the fuel cell product. However, if you are of a more cynical bent, you may also read into this as a delaying tactic.

The internal combustion engine fuelled by hydrogen is also a tantalising option and one that could pose more of a threat to fuel cells than do hybrid technologies. The consumer, when adopting one of these cars, will only have to switch fuels, something that we have historically seen is possible with enough price incentive. BMW has released plans to put dual-fuel 7 series vehicles on the road in two years (2009) time. Mazda is already leasing its version of a hydrogen ICE. Ford is also very active in this area but so far has given no dates for a commercial product.

With the development of high-pressure hydrogen storage tanks, the main barrier to adoption of these vehicles now is the refuelling question, and this is being addressed across a number of countries.

Taking all of this technological activity into account, it should be remembered that one of the main deciding factors for purchase choice is vehicle appearance. The designs that are now starting to appear for FCVs, including the FCX already shown and the GM Sequel, indicates that FCVs might actually win on looks.

2.6 Summary

The aim of this chapter has been to provide a basic introduction to fuel cell technology and a number of other application areas not covered in the remainder of this book. The main points to take away are that there are a number of fuel cell stack technologies under development, each with applications areas that are more suited to than others. To wait to see if one type becomes dominant in an application closing the market for others will take a long time as to date there are no signs of this happening. Apart from the main six types we may see some of the less mainstream fuel cell types breakthrough into a more commercially orientated phase.

Further reading

There are a growing number of detailed engineering tomes on fuel cell technology, and if the reader now has an interest to find out greater technical detail two further books that can be recommended are:

1. Singhal, S. and Kendall, K., 2003, High Temperature Solid Oxide Fuel Cells: Fundamentals, Design and Applications, published by Elsevier (ISBN: 1-85617-387-9)
2. Vielstich, W. Lamm, A. and Gasteiger, HA., 2003, Handbook of Fuel Cells: Fundamentals Technology and Applications (volumes 1–4), published by Wiley (ISBN: 0-471-49926-9)
3. Dicks, A. Larminie, J.C., 2003, Fuel Cell Systems Explained, published by Wiley (ISBN: 047084857X)
4. Basu, S., 2007, Fuel Cell Science and Technology, published by Springer-Verlag New York (ISBN: 0387355375)

3
Drivers for stationary fuel cells

3.1 Introduction

Access to power is one of the foundations of modern society. The current paradigm in which we operate provides electricity via a centralised model. Power is produced from a variety of fuel sources such as coal, nuclear and renewables and also uses fossil fuels, such as petroleum, in various conversion technologies such as the internal combustion engine and generators. Both of these energy strands are now embedded into everyday life in developed and increasingly in developing countries.

Consumption of this energy has steadily increased over time and is predicted to continue to increase. Figure 3.1, developed from data from the International Energy Agency (IEA), shows the historical and projected world energy consumption.

The world has entered a period of political, social and technology where a number of issues are being raised with the energy systems that we operate in. These centre around two foundation foci – climate change and energy supply.

Both of these two fundamental issues are key drivers for change, not just in terms of how our energy is produced but also the efficiency of the way we use our energy and also the forms of energy we use. Fuel cells, with their inherent efficiency gains and ability to use a range of fuels, including hydrogen, are being driven to market by both of these factors. Each of these issues is examined in more detail below, with some application-specific drivers also being covered in relevant chapters later.

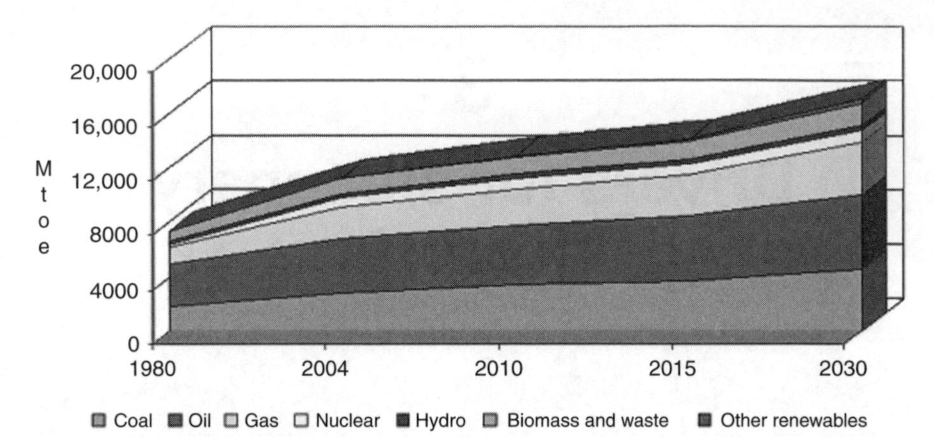

Figure 3.1 Projected world energy consumption by fuel type
Source: World Energy Outlook © OECD/IEA, 2006, Table 2.1, p66, as modified by the Author

3.1.1 Climate change

The current political debate on our impact on the natural environment began with the publication of the Brundtland Commission Report 'Our Common Future' in 1987. Within this report, its author, former Norwegian Prime Minister Gro Harlem Brundtland, coined the concept of *sustainable development*.

> Our common future definition of sustainable development:
>
> Sustainable Development is development that meets the needs of the present without compromising the ability of future generations to meet their own needs

Since the publication of this seminal report, a level of consensus has been reached that we, through the unprecedented release of greenhouse gases into the atmosphere, are causing climate change. Though this statement can cause something of a reaction of 'well we have known that for some time', it has not been until the publication of two very recent studies that appears to have, finally, caused international governmental agreement that the time for action not only to mitigate, but also to work on living within the effects of climate change, is now.

The Stern Report

The Stern Report, written by the prominent British economist Sir Nicholas Stern, was published by the UK government in 2006. The report, and the research behind it, was a thorough examination of the global economic impacts of climate change and opens with the chilling statement:

> 'Climate change presents a unique challenge for economics: it is the greatest and widest-ranging market failure ever seen'.

It goes on to say:

> '...the evidence gathered by the Review leads to a simple conclusion: the benefits of strong, early action considerably outweigh the costs'.

The report analyses the period of change that we are facing in the transition to a low-carbon economy and what governments, and industry, can do now.

The conclusion of the report is that we are facing a window of opportunity where if we take action now we can not only mitigate the worst impacts of climate change, by stabilising the levels of carbon di-oxide emissions in the atmosphere, but also by using market mechanisms such as carbon trading we can decrease the costs of adoption and increase the speed of diffusion of a range of low-carbon energy technologies. If, on the other hand, we choose to do nothing (the so-called business-as-usual scenario), then:

> '...Business As Usual climate change will reduce welfare by an amount equivalent to a reduction in consumption per head of between 5% and 20%'.[1]

The 4th International Panel on Climate Change (IPCC)

The IPCC is a non-governmental group of independent scientists working together to research the causes and impacts of global climate change. Though this group has been criticised by some as being a political pawn and has experienced a degree of infighting, the work of the IPCC is widely regarded

[1] As with any other report, this publication is not without its criticism. The main criticism that has been levelled against it is the choice of a near-zero discount rate and not the current market-based rate.

as the benchmark upon which climate change-related governmental action is based. The latest report, published in February 2007, stated that:

> 'Most of the observed increase in globally averaged temperatures since the mid-20th century is *very likely*[2] due to the observed increase in anthropogenic greenhouse gas concentrations'.

and

> 'The observed widespread warming of the atmosphere and ocean, together with ice mass loss, support the conclusion that it is *extremely unlikely*[3] that global climate change of the past fifty years can be explained without external forcing, and *very likely* that it is not due to known natural causes alone'.

Another thread throughout the report was the impact of climate change on global weather patterns. Though there has been some controversy over linking the increasing occurence of hurricanes to climate change, there appears to have been consensus amongst IPCC members on the increase and frequency of extreme weather patters such as drought and flood and heat waves and extreme cold. It is these second set of extremes that could see an increase in demand for energy, in terms of space heating and cooling, whilst the first will push up the demand for secure power supplies. Both of these are examined in more detail below.

3.2 Why climate change is a driver for stationary fuel cells

The increased politicisation of the environment and climate change has led to a call from most world governments for a substantial reduction in the amount of greenhouse gases emitted into the atmosphere. Energy, through the power sector, and use in buildings are two major contributors to greenhouse gas emissions, as seen in Figure 3.2.

A number of industries, to a greater or lesser extent, have taken up this challenge of emission reductions and are working on technologies with higher efficiencies of conversion and technologies that are inherently cleaner

[2] *very likely* = 90% probability.
[3] *extremely unlikely* =< 5% probability.

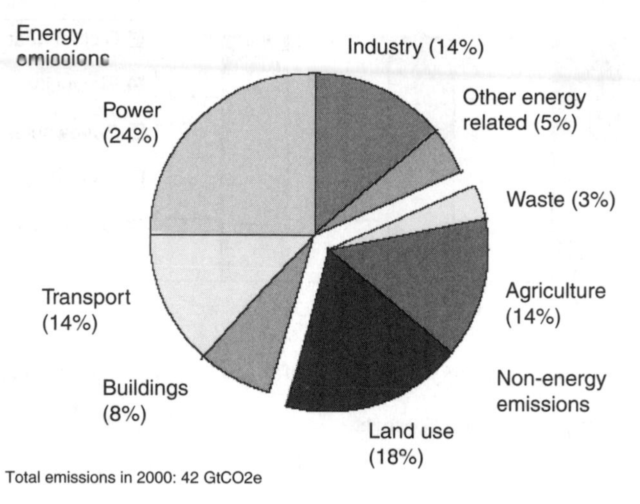

Figure 3.2 Greenhouse gas emissions by sector
Source: The Stern Report, © Crown Copyright

due to low-carbon fuel use. Fuel cell technology, with its high efficiency and potential to use utilise low- or zero-carbon fuel sources, fits neatly in with both strands under development.

3.2.1 Increased efficiency of energy conversion

According to the work carried out by the IEA, and used as evidence in the Stern Report, buildings currently emit approximately 8% of greenhouse gas emissions directly. When the IEA's model assumptions also take into account upstream pollution from the electricity production, or fuel extraction and processing, then this number jumps to 20%. This equates to 3.3 GigaTonnes (Gt) of carbon di-oxide from direct combustion of fossil fuels in residential and commercial buildings, and over 8 Gt of carbon di-oxide when the upstream portion is also taken into account.

If this trend continues under a business-as-usual scenario, we could see this increase to over 20 Gt carbon di-oxide in 2050. This is shown in Figure 3.2. Note that both Figures 3.3 and 3.4 have been taken from the Stern Report and are based on the scenario work done for this.

It is already clear though that due to increased government and industry action, this business-as-usual scenario will not be allowed to happen. Recent developments include a variety of policy measures towards creating low- or zero-carbon buildings. A number of these policies are outlined in Chapter 7.

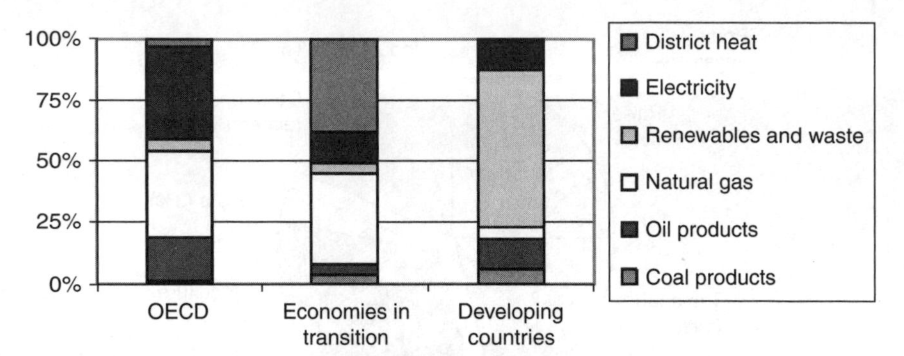

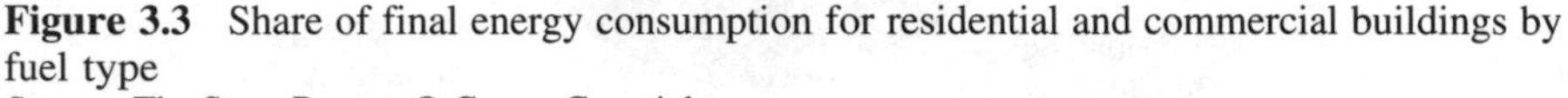

Figure 3.3 Share of final energy consumption for residential and commercial buildings by fuel type
Source: The Stern Report, © Crown Copyright

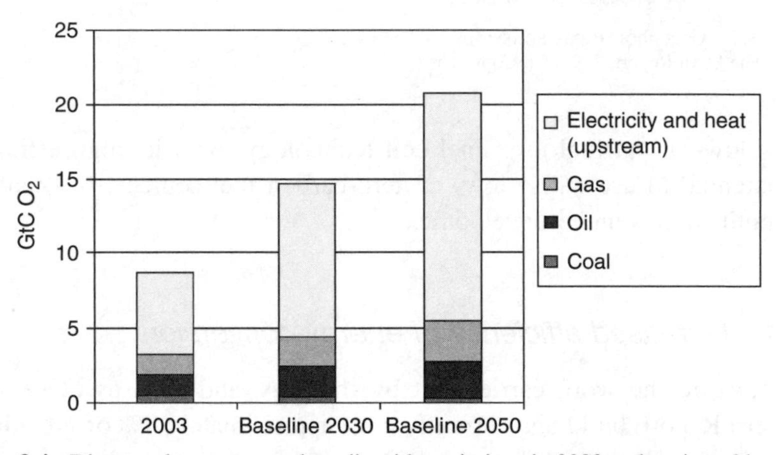

Figure 3.4 Direct and upstream carbon di-oxide emissions in 2003 and projected business-as-usual emissions for 2030 and 2050
Source: The Stern Report, © Crown Copyright

Within this policy debate, fuel cells are seen as being one of the key technologies that will facilitate the transition towards a low-carbon economy. Whether they are deployed in the home, the office or as uninterruptible power supply (UPS) systems, what we are likely to see is a growing proliferation of these units over the coming decades.

3.3 Security of supply

3.3.1 Government

Concerns over energy supply are becoming increasingly politically sensitive and are centred around the question of how long will we have easy (or cheap) access to fossil fuels. Note that this question is not the same as when will oil run out – a debate that is raging seemingly without end. Oil and also natural gas are finite resources located in a limited number of geographical locations. Whereas the debate of when the oil wells will run dry seems to be looking at a longer term picture, the problem that does exist is where these wells are. Figure 3.5, using data from the 2006 BP Statistical Energy Review, shows the location of current oil reserves and resources, whilst Figure 3.6 shows the same for natural gas.

Though at present a non-dominant percentage of petroleum reserves lie within OPEC (oil-producing and exporting countries), current projections indicate that within the next two decades this could switch with OPEC containing a majority of oil reserves. (Figure 3.7)

Looking to the future we have a number of options, some technological and some consumer based, to reduce the impact of the energy systems and alleviate concerns over energy supply. Put very simply, one of the main technological options that we have is to increase the efficiency of the conversion process from petroleum to power, or fuel to electricity.

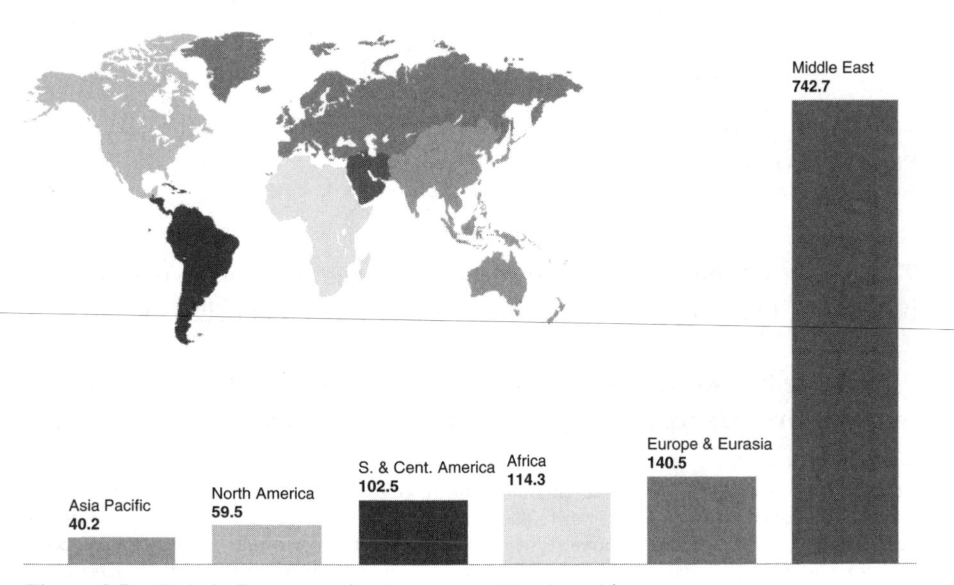

Figure 3.5 Global oil reserves (in thousand million barrels)
Source: BP Statistical Review of World Energy 2005, © BP

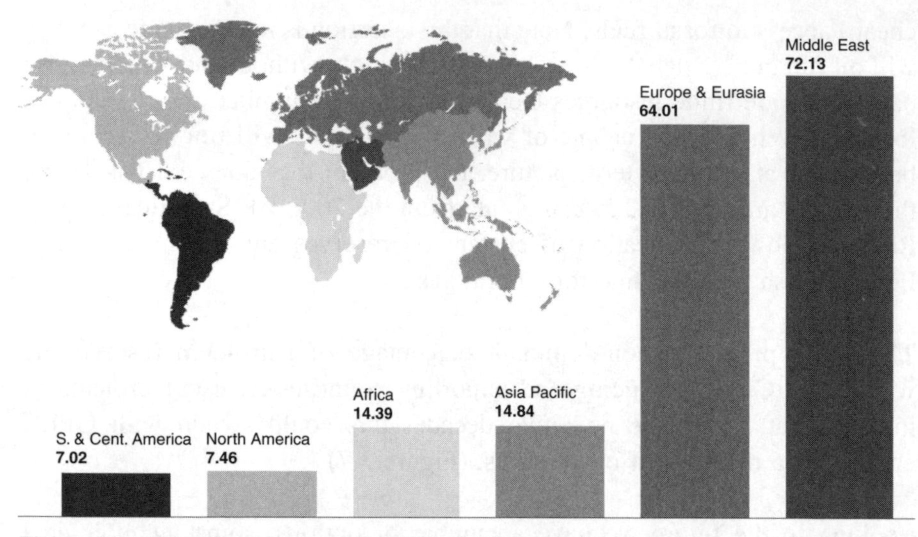

Figure 3.6 Natural gas reserves (in trillion cubic meters)
Source: BP Statistical Review of World Energy 2005, © BP

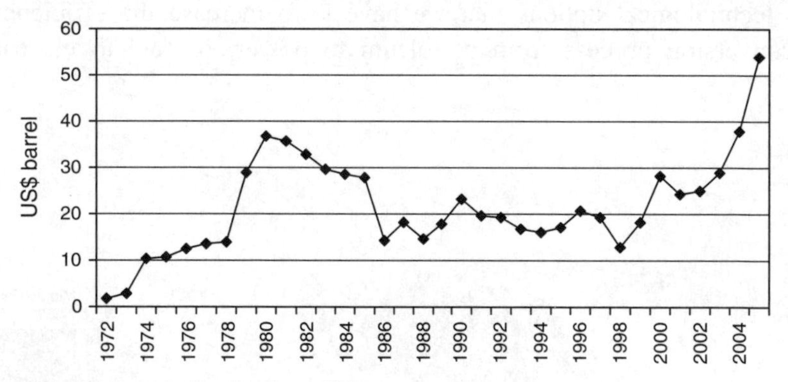

Figure 3.7 Historical oil price from 1960
Source: Modified from BP Statistical Review of World Energy 2005, © BP

The impact of this would be to reduce both the amount of fuel in and, potentially, the level of pollution out. This is where fuel cells come in.[4]

As fuel cells use hydrogen, which can be produced from a range of locally derived sources, most governments see this, linked in with increased

[4] The emissions profile of a fuel cell depends on the fuel used to produce it and efficiency of the system under examination. There are cases where the emissions can be zero and cases where the emissions can actually be higher than the current systems in place. As it is not within the remit of this book to look into hydrogen production this is all that will be said on this area except that in most cases using hydrogen in a fuel cell is more efficient and produces fewer emissions than the current suite of incumbent technologies.

energy efficiency, as one of the methods of reducing the dependence on imported oil.

3.3.2 Consumer

Along with the governmental level of drivers of climate change and energy conservation security, the third main driver comes at adopter level, namely that of security of access to power.

In the UPS/backup power sector, for example, the reason for the interest in this application is that any downtime obviously has financial impacts on a company, with this cost naturally increasing the length of time the equipment is off-line. In terms of these costs to industry quoted cost figures from GoldenGate Software (2005) are summarised in Table 3.1.

What is not included in these calculations are less quantitative costs such as damage to corporate image and investor confidence and knock-on effect on customers and suppliers.

As more companies become aware of these costs and develop internal disaster recovery scenarios, UPS technologies are also becoming more mainstream and cost-effective. Utility spikes and sags are known to be a fairly common occurrence and need high-grade electricity to smooth them out, but it is the brown- and blackouts that grab the headlines. Hurricane Katrina, which hit the Eastern seaboard of the USA in 2005 is calculated to have done US$20 billion worth of damage due to the region's loss of power. Other high-profile grid failures in the US were the rolling blackouts that California experienced during 2001 and the North East blackout in 2003, which affected 50 million people across eight US States and the Canadian province of Ontario.

Table 3.1 Average cost of downtime

Industry	**Average cost of downtime* (US$/h)**
Mobile phone company	41,000
Telephone ticket sales	72,000
airline reservations	90,000
Credit card operations	2,500,000
Brokerage operations	6,500,000

*Note downtime can be significantly less than recovery time

Europe also suffers from brown- and blackouts, with the largest, to date, being the 2003 shutdown in Italy, which affected an estimated 53 million people.

These large-scale, attention-grabbing incidents are still thankfully fairly rare, but with an expected increase in severe weather patterns, which some link with climate change, catastrophic utility shutdown scenarios are predicted to increase as well.

Short-term blackouts, for a period of hours instead of days, are more common, especially in Europe. These can occur from something as trivial as the extra power drain from the millions of kettles being flicked on during a break in a high-profile football game to the power lines being torn down in a localised storm.

All this adds up to a growing market for products that will provide reliable and measurable power, in terms of electricity, in cases of utility fluctuation, brown- or blackout.

Other market-based stationary drivers include:

- High-temperature water by-product – for heating in various applications
- Mitigating or delaying the need for grid reinforcement in decentralised generation formats
- The ability for accurate load-following (both scale-up and scale-down) made possible by the highly modular nature of many types of stationary fuel cells
- The possibility of avoiding electricity wires or gas/oil pipes to stationary loads in remote applications

As these drivers are examined further in upcoming chapters only a brief mention of them has been made here.

3.4 Summary

When the modern fuel cell movement was first starting to become established, the common cry was 'fuel cells will be a success because they are good for the environment'. This singular vision has steadily been replaced with a broader basket of drivers, still including a desire to reduce the impact on the environment, but also now including a number of market- and regulatory-based ones.

4

Fuel cells for the home

4.1 Introduction

A range of publications, from those targeted more at the survivalist culture to the mainstream architectural trade press, are increasingly covering fuel cells as a medium-term option for residential power. Now with governments also waking up to the potential of the residential fuel cell, especially in terms of the drive to the low- or zero-carbon home, the pressures for change are aligning to present a window of opportunity for this application.

This chapter looks at the two main options under development: micro-combined heat and power (mCHP), which is grid supported, and off-grid power. It discusses the speed of development in each, the main developers, costs, fuel choice and also provides a short section on policy. Note that the main discussion on policy is in Chapter 7.

4.2 Development of residential fuel cells to date

In terms of the number of demonstration units deployed, residential fuel cells have nearly doubled over the past three years. This trend is expected to continue well into the next century. Figure 4.1 shows this growth trend, whilst Figure 4.2 puts these developments in context to all other fuel cell applications. What can be seen is that at the time of this writing, residential fuel cells represent around one-fifth of all total installed fuel cells. Note that this (or what) is cumulative over this period time.

The reason for the growth in units is twofold. First, a number of major utility companies are operating beta test trials of 100 plus units each. These

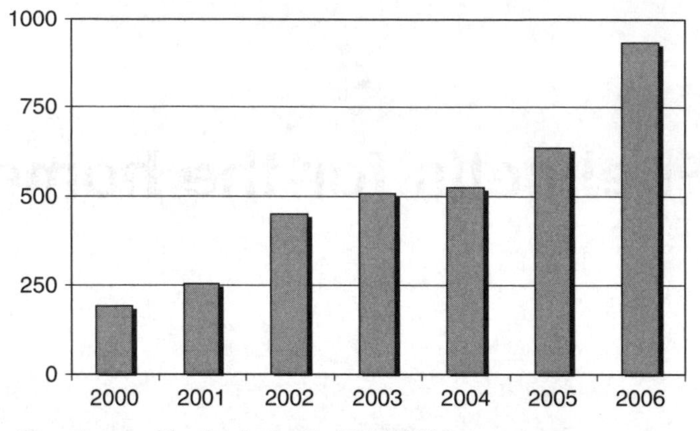

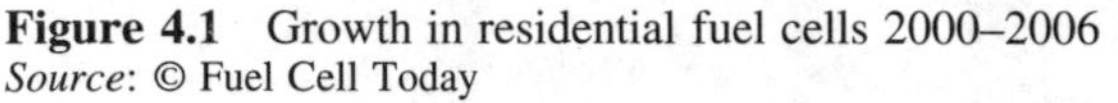

Figure 4.1 Growth in residential fuel cells 2000–2006
Source: © Fuel Cell Today

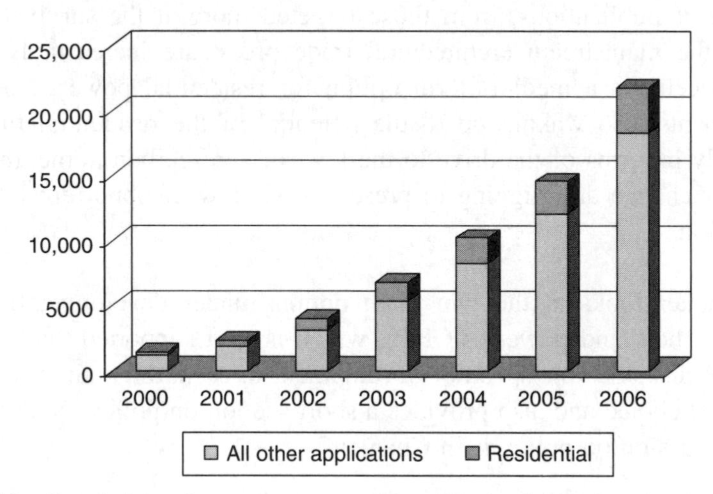

Figure 4.2 Cumulative units, as a percentage of total fuel cells developed
Source: © Fuel Cell Today

trials are centred mainly in Europe and Japan.[1] Second, government policy is increasingly removing barriers to adoption of the units. This second part is non-trivial as it was not until the first literal in-house testing was done that a number of the barriers, such as location of the fuel cell, venting and if needed access to hydrogen storage areas, were able to be identified.

[1] Historically in the USA, the military has undertaken substantial fuel cell demo programmes, including testing of a number of units for the home. These tests have now been completed and little in the way of residential level testing is *currently* underway in the USA.

As with uninterruptible power supply (UPS) and power plant fuel cells, there is no 'one-size-fits-all' unit being developed. To date, the most common sizes, in terms of units developed and tested, are 1, 1.5, 3, 5 and 10 kW. In general, the lower the kilowattage the less time the fuel cell unit acts as the primary power source. For example, many of the 1 kW units, which are primarily being developed in Japan by firms such as Tokyo Gas, Nippon Oil and Toyota, are providing base load power for 8 hours during the day and the heating for hot water requirements. The rest of the time the electricity for appliances and space heating is drawn down from the grid as normal. This mCHP option is looked at in more detail in section 4.3. It is not until you realistically get past 5 kW that you can think of using a fuel cell to cover power requirements in an off-grid situation.

The main advantages of fuel cells for residential applications are the same as for many other applications but also include removing strain from the electricity grid, which is a utility-level benefit, and reduction in household heating bills due to the waste heat being used to heat the hot water tank, which is a consumer benefit.

The current main disadvantages of using a fuel cell for residential applications include the cost, maintenance requirements and diverse connection issues. The first two, cost and maintenance schedule, are expected to decrease as more units are produced and as stack technology evolves, whereas connection issues are down to industry, and politicians to sort out a more open marketplace that allows for distributed generation (of all sizes).

At present, depending on which country, you live in hurdles such as actually procuring the units, grid interconnection issues, feed-in tariffs and monitoring of the fuel cell make planning a fuel cell-powered home somewhat challenging.

Grid interconnection

Grid interconnection is the set of rules and regulations allowing electricity-generating equipment to be hooked up and into the electricity grid. Different areas have different regulations, but the overarching theme is to ensure security and stability of the electricity grid. In the USA, these regulations are set at state level, in Europe they are set, currently, at national level (though this appears to be changing) and in Japan it is set at national level. Though a growing number of manufacturers are having their product lines approved for grid interconnection in different regions, the industry is far from having a broad coverage so each installation at present has to work on something of a case-by-case basis.

Feed-in tariffs

A feed-in tariff is the price that is paid for electricity that is fed into an electricity grid. In theory, if a grid-connected fuel cell produced power more than was being used by the household it can be fed into the grid. Other renewable energy sources that have already gone down this route have developed a system of net metering to allow the customer to monitor how much electricity has been fed back into the grid and how much credit is due. Not all countries have feed-in tariffs, and in some this lack of grid accessibility can be a barrier to residential fuel cell deployment.

The next section of the chapter splits the types of development into the two main groupings – grid supported (mCHP) and off-grid fuel cells.

4.3 Micro-combined heat and power

The use of CHP, where the waste heat from power generation is captured and used either as direct space heating or to heat or cool water, is not new but until recently has been marginal in many countries in terms of installed capacity. More recently, though, interest in CHP has significantly increased. Specific drivers for this movement include grid reliability issues, the need by utilities to reduce emissions through more efficient power generation, shown in Figure 4.3, and the decreasing cost of CHP options. To date, most CHP units have been in the low tens of kW's but now though with the development of new technologies mCHP is coming to the fore.

mCHP, which is defined by CoGen Europe as any unit under 3 kWe, is aimed squarely at single home buildings. Unlike mini-CHP, defined as any unit under 100 kWe, which can be units used to power a block of flats, for example, or a series of houses (for the purpose of this book these types of units are looked at in Chapter 5), mCHP technology has only a limited number of options.

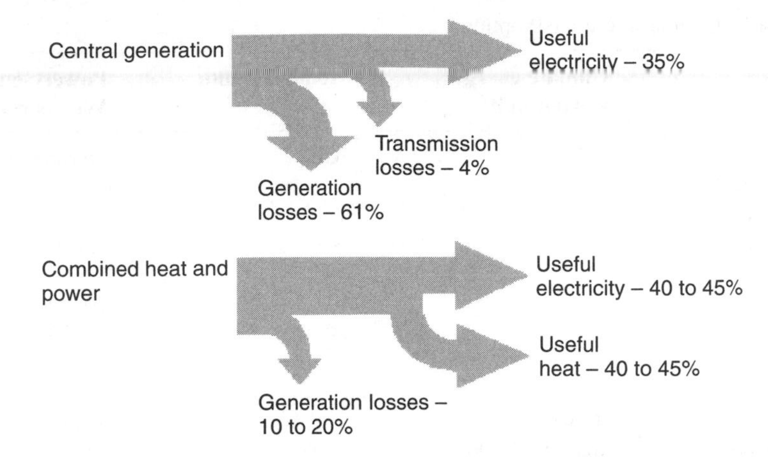

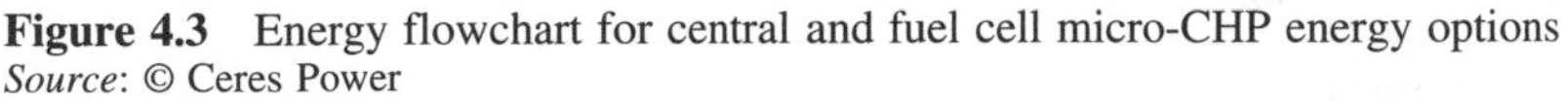

Figure 4.3 Energy flowchart for central and fuel cell micro-CHP energy options
Source: © Ceres Power

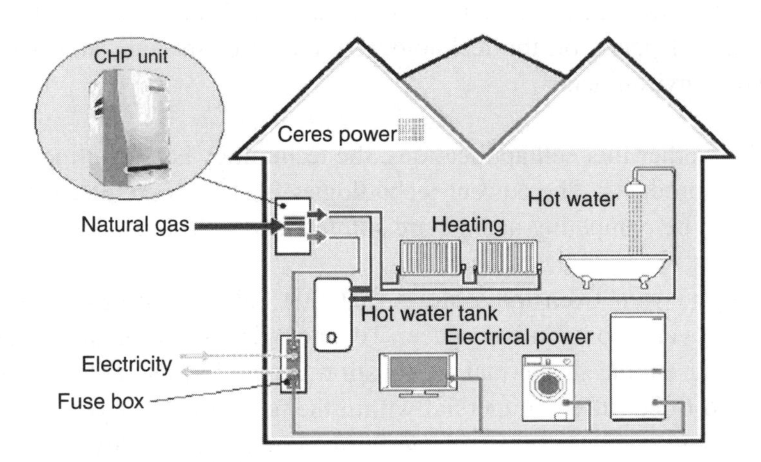

Figure 4.4 A Fuel cell CHP home
Source: © Ceres Power

Normally, an mCHP-powered home would have a small (usually) refrigerator-sized unit located somewhere in the house[2] which provides electricity to the fuse box and heat for either the hot water tank, or less often, space heating. Figure 4.4 shows an envisaged typical fuel cell mCHP home.

[2] The actual location of the unit varies from country to country – in Europe and the USA they are usually located indoors, but in Japan they are always located outside.

Table 4.1 Current micro-CHP options

	Climate energy CoGen unit	**Honda IC unit**	**PowerGen/ WhisperGen**
Technology	ICE	ICE	Sterling engine
output (kW)			
Electrical	1.2	1	1
Thermal	3.2	~3	7.5–12
Efficiency (%)			
Thermal	65	65	70–80
Electrical	20	20	10–20
Combined	85	85	90
Emissions			
CO_2	'Lower'	<30%	<20%
Cost (US$)	10,000–12,000	7500	5500

Source: Product literature

In terms of how much you can get out of a mCHP unit, the actual heat to power ratio depends on the technology, the size of the unit and the local regulatory environment.[3]

As with all other fuel cell applications, the technology is competing directly with other products. The current technologies in this marketplace that fuel cells would be competing against are summarised in Table 4.1.

What can be seen from the table is that fuel cell technology is entering an already very competitive space. Technologies such as the Whisper-Gen sterling engine (a thermal-combustion generation technology), which is expected to be fully commercial within the next couple of years, already have excellent conversion efficiencies and a comparatively low price tag.

For fuel cells to succeed in this space, they will have to provide added extras to encourage adoption by the consumers and support from the utilities. These added extras could be potential zero greenhouse gas emission operation, the fact that they are silent or, if they are, government subsidies for adoption. In summary, these non-technological issues to be worked out are still significant.

[3] In a number of countries, it is illegal to dump waste heat into the external environment so an mCHP unit has to be sized correctly as not to produce more heat than can be used by the household at any given time.

4.3.1 Fuel cell technology under development

Looking solely at fuel cells under development, for residential applications two types are currently being worked upon: proton-exchange membrane fuel cells (PEMFCs) and solid oxide fuel cells (SOFCs). As has been outlined in Chapter 1, PEMFCs are low-temperature units, normally operating under 90°C, and SOFCs are high-temperature units, normally operating over 800°C.

4.3.2 Main companies involved in fuel cell mCHP

PEM-based units

The number of companies operating in the PEM-based residential fuel cell space is limited. Three of the main players in this sector are:

Ballard (Canada, http://www.ballard.com) is arguably one of the world's most famous fuel cell companies. Created in 1979, Ballard came to the fore in the 1990s with its breakthrough PEM technology. Now in conjunction with its Japanese joint venture company Ebara Ballard, it has released its latest residential co-generation fuel cell. The 'Mark 1030 V3' 1 kW stack has been designed to work towards the 2008 Japanese government targets of 10 years' lifetime in the home (equivalent to 40,000 hours operation).[4] The stack is 40% lighter and 26% smaller than the earlier V2 stack. Development aims are now to further reduce the costs of the stack and of the fully integrated unit. In terms of marketing, Ebara Ballard has exclusive rights to the CHP units in Japan, whilst Ballard holds the rights for marketing to the rest of the world (Figures 4.5, 4.6).

Though Ballard has released a number of technology-based roadmaps focusing on meeting the automotive targets, set by the US Department of Energy, and is arguably better known for its work with fuel cell vehicles, as a company it has also made it clear that it sees the stationary fuel cell market as being a key element of its future growth.

Baxi Innovation, formerly European Fuel Cells (Germany), owned by Britain's Baxi Group, is developing and producing residential PEMFC units. The company's natural gas-powered 1.5 kW beta unit is currently being field tested. It is developing what it terms a fuel cell heating unit (FCHU), which

[4] A full English translation of the 2006 NEDO Fuel Cell Roadmap which includes a full version of the 2008 residential targets can be found at Fuel Cell Today: http://www.fuelcelltoday.com/FuelCellToday/FCTFiles/FCTArticleFiles/Article_1131_2006%20Japanese%20Roadmap%20-%20English%20Translation.pdf.

Figure 4.5 Three generations of ballard fuel cell stacks
Source: © Ballard

Figure 4.6 Three generations of ballard fuel cell stacks
Source: © Ballard

will cover 75% of the heating demand of a typical European single family home. The company has signed a co-operation agreement with IRD Fuel Cell for development and production of fuel cells and system components. In terms of full commercialisation, Baxi gives target dates of 2012–2015, with prototypes available now. For end-user consumers, the aim is that the units will be on sale via standard high-street salesrooms and will be marketed as new boilers that also produce electricity.

Toshiba Fuel Cell Power Systems (TFCPS) (Japan) a subsidiary of Toshiba, has the sole purpose of commercialisation of its 1 kW residential PEMFCs by 2008. This is termed the 'Dash to 2008' when it is planning on having a unit priced on the open market at less than 1.2 million Yen (approximately US$9500). The technical targets for the system are a cold start time of less than 10 minutes, overall efficiency of >77% higher heating value (HHV) and 80°C waste heat. In the new company, TFCPS will produce the stacks,

Toshiba Home Techno assemble will the systems and the reformers will be sourced externally.

The current unit, the 'TMI-A', is designed to be operated in a daily start–stop mode and provides the residence with all of its hot water and most of the base load electricity requirement during the day. The current unit has a total efficiency of 71% (HHV) and can be run of LPG or natural gas. The unit is not fuel flexible and cannot be switched between fuels.

Note that TFCPS has yet to have any distribution agents outside Asia.

SOFC-based units

Small SOFC development is comparatively recent and has not yet had the level of investment that PEM has attracted. Less than a dozen companies globally are currently working on SOFC residential units. Four of the main players are listed below.

Ceramic Fuel Cells (Australia and the UK) is a manufacturer of 1 kW SOFC units. Its 'NetGen' unit is designed to operate continuously, thereby operating primarily as an electricity generator, with an overall target efficiency of >80%. The unit also has an inbuilt integrated reformer. Currently, it can be fuelled with natural gas, but the future targets include LPG, propane, biogas, ethanol and hydrogen.

Ceramic claims to have a unit suitable for large-scale manufacturing with important reported metrics including increased power density up to 400 mW/cm^2, electrical conversion efficiency of 50% and fuel utilisation up to 85%.

Looking at manufacturing, we are expecting to hear an announcement soon (2007) as to the location, within Europe, of Ceramic's first commercial-scale manufacturing facility. Also Ceramic Fuel Cells is one of the few non-Japanese companies to demonstrate under the NEDO residential fuel cell programme. As of FY2007, Ceramic will have a number of its SOFC systems being used and monitored for a period of 4 years within this programme.

Ceres Power (UK), an SOFC stack manufacturer, has signed an agreement with British Gas, funded by the UK Department of Trade and Industry (DTI), to develop and market SOFC residential fuel cell mCHP units. The SOFC stacks being developed by Ceres operate at a lower temperature than on average due to a combination of ceramics and stainless steel. The deal with British Gas is to develop a unit specifically for the UK market, with power output of around 4 kW. Ceres is following the same route as a number of other companies working in the residential space by exploring tie-ups with boiler manufacturers, allowing them reciprocal access to stack technology and know-how.

In terms of manufacturing capability, 2007 will see the commissioning of a pilot scale facility, to be up and running by mid-2007, and the start of the design for the scaled-up full size manufacturing facility. It is anticipated that this second facility will be fully commissioned during 2008.

Kyocera (Japan) is also developing 1 kW SOFCs for residential use and larger units for businesses.

4.3.3 Economics

As this application is still very much in the beta testing phase and is some years away from being commercial, current costs are not only difficult to track down but also difficult to extrapolate forward. Most of the current costs, which are summarised in Table 4.2, come from Japan, which has significant experience in residential demonstration programmes.

What can be seen from Table 4.2 is that costs, as mentioned before, are still prohibitively high for the mass-market consumer. Also as costs in Japan are artificially 'low', due to the high level of government subsidies, the actual

Table 4.2 Current (2005–2007) residential fuel cell costs

In-use costs	Location	Technology	Size (kW)	Cost (US$)	
				Buy/kW	Lease
Reference	Japan	PEM	1	80,000	
Cosmo Oil	Japan	PEM	0.7		500*
Toshiba Fuel Cell Power Systems	Japan	PEM	1		840*

Source: Fuel Cell Today.

* part of the Japanese Stationary Fuel Cell Demonstration Programme and receiving subsidies from the Japanese government.

Note that as with any other part of the fuel cell industry, a large number of reports, with modelled costs, have been published in this area. There has been a conscious decision not to include this information here due to a need to report all the assumptions etc that go behind the output costs. A number of a these reports are listed at the end of the chapter and can easily be accessed if required.

Table 4.3 Future/projected residential fuel cell costs

Manufacturer	Region	Date	Technology	Cost (US$/kW)
Hokkaido Gas	Japan	2008	PEM	8000
Ebara Ballard				
Natural Gas	Japan	2008	PEM	10,000
Kerosene	Japan	?	PEM	12,000
Fuji Electric	Japan	2008	PEM	12,000–16,000
Advanced Technology	Japan	2015	PEM	2500–4000
Matsushita Electric industrial	Japan	2008	PEM	10,000
Toshiba Fuel Cell Power Systems	Japan	2008	PEM	9500

Source: Fuel Cell Today.

Note that as with any other part of the fuel cell industry, a large number of reports, with modelled costs, have been published in this area. There has been a conscious decision not to include this information here due to a need to report all the assumptions that go behind the output costs. A number of a these reports are listed at the end of the chapter and can easily be accessed if required.

cost per unit to the manufacturer is unclear. It should be kept in mind though that it is still very early days for this application, and costs are projected to quickly come down over the next decade.

Looking to the future, Table 4.3 tabulates a number of manufacturer's target costs and timetable.

4.4 Off-grid power

A recent UN report states that globally over two billion people do not have access to any electricity. Any technology that can help to alleviate this problem will provide major societal benefits. Because of this issue, off-grid power has to be looked at as a potentially major market for fuel cell technology.

Looking at technology matching, what the market needs for off-grid power is far more challenging for fuel cells to provide than for mCHP. Not only do they have to be capable of providing base load power at all times but also be able to cover demand spikes. A recent analysis done by Ceres Power, shown in Figure 4.7, shows that the base load power is actually fairly low for a typical UK home (the model was UK based), but the spikes represent a sudden surge of nearly triple the base load.

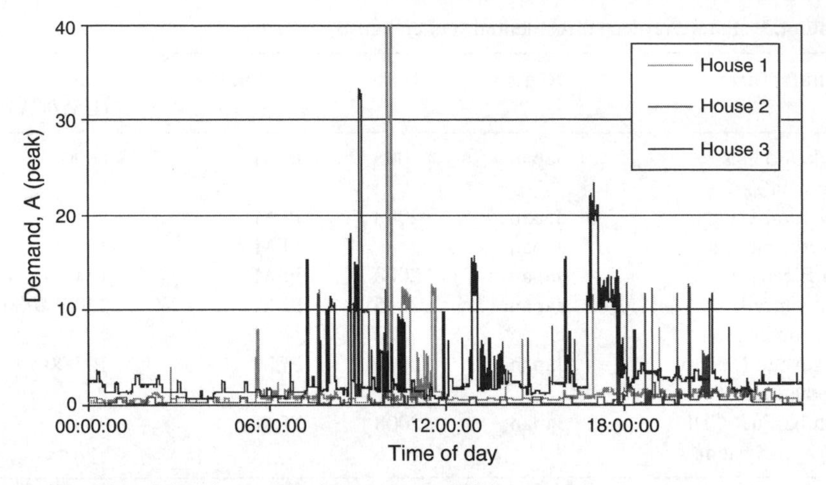

Figure 4.7 Electricity demand in three average UK homes
Source: © Ceres power

So even though the base load may be in the low kilowatts, if the peak load is much higher, then the unit would need to be scaled to, and therefore be capable of, meeting such instantaneous demand.[5]

Due to this load requirement issue, most units, of which there are few that are currently being designed to fit into this application, are around 10 kW, or conversely 2 kW. For the 10 kW unit, around 5 kW of this is for base load power with the other 5 kW being available to meet peak load power. The smaller unit works on the premise that the house has a range of other energy technologies[6] in place, so that the fuel cell is not there to provide the majority of electricity and heating requirements, but is one of a range of complementary technologies.

An added complication, if looking at procuring units for an off-grid house, is that most of the development of smaller residential units (<10 kW) is being done through tie-ups with fuel utilities, such as the natural gas companies. The gas company's role is commonly to provide a market entry route, such as boiler replacement, as a known and trusted brand as well as supplying a fitting and maintenance team. The number of fuel cell companies that are looking at developing a fully integrated product rather than licensing

[5] One issue is how much of this can be mitigated by consumer education? If peak loads are created by the usage of a number of pieces of electricity-hungry equipment then educating the consumer to simply not switch them all on at the same time may help to provide a partial solution.

[6] Including highly efficient electrical equipment and insulation.

its stack technology with distribution channels that could be used for this off-grid market is very limited.

If a project was seriously looking to develop a single fully off-grid home, it would be more advisable to look at the fuel cell companies that can sell, through a distribution agent, a small unit direct and working with a project developer to use the fuel cell as the house's power source. All said and done, at present, this is a very challenging route to go down.

4.4.1 Technology

As with grid parallel developments, the main two technologies being looked at are PEMFCs and SOFCs, as these can most readily be scaled to match demand.

4.4.2 Main companies involved in residential off-grid power

Ceres Power (UK), working with British Gas and also in partnership with Linde, states that this relationship is to develop off-grid generators. Though little else is currently known about this market strand for Ceres, what is known is that the unit under development will function as a CHP unit and could be fuelled using bottled liquid petroleum gas (LPG).

IdaTech (USA) is one of only two companies listed here that can provide an off-grid fuel cell package to the market now. Its 2.4 kW hybrid system (1.4 kW PEMFC and a 1 kW solar array) is linked in with a solar panel for rural locations. Though primarily intended to provide power to telecommunications, the unit does produce useable heat, potentially allowing it be used as an off-grid mCHP unit.

Intelligent Energy (UK) has also made a clear statement that they are working on premium distributed power for residential, industrial and commercial sectors. Its PEM technology has recently been showcased in small rural applications in South Africa with great success. Though these were not powering buildings, but critical equipment, it has shown that using fuel cell technology is viable in some off-grid applications now.

Plug Power (USA) is producing a range of PEM-based products for the UPS market. Included in these is a unit for primary off-grid power. A recent financial filing of the company did state that it was also developing 'additional products, including a continuous power product, with optional

CHP capability for remote small commercial and remote residential applications'. The unit is being designed to run on a number of different fuels.

4.5 Fuel choice

Fuel choice is not as much of an issue for residential fuel cells as it is for, say, transport. To date, most of the units being tested are using hydrogen-rich sources such as natural gas and LPG. Figure 4.8 shows the fuel mix that was used during 2006 in residential units.

Of all the fuels listed, kerosene constitutes the biggest challenge in terms of fuel reforming. Depending on the type of fuel cell being used, the level of impurities has to match to its tolerance level. PEMFCs historically have required very high-purity hydrogen, in the region of 99.95% pure, whilst SOFCs are more tolerant and can use comparatively 'dirty' hydrogen.

In an ideal situation, the choice of fuel cell would depend on the house's location and the price of the fuels. If a house is already connected up to a natural gas grid and the price differential between electricity and natural gas is great enough to encourage usage of natural gas, then in economic terms the use of natural gas is valid. If the house though is not hooked up to the electricity grid or to the natural gas grid, then either using tanked in LPG, which can be stored on site, or ideally, local renewables could be a more economic route.

What we expect to see as this market develops further is a definite push from the fringes of off-grid homes, either in rural or in urban situations,

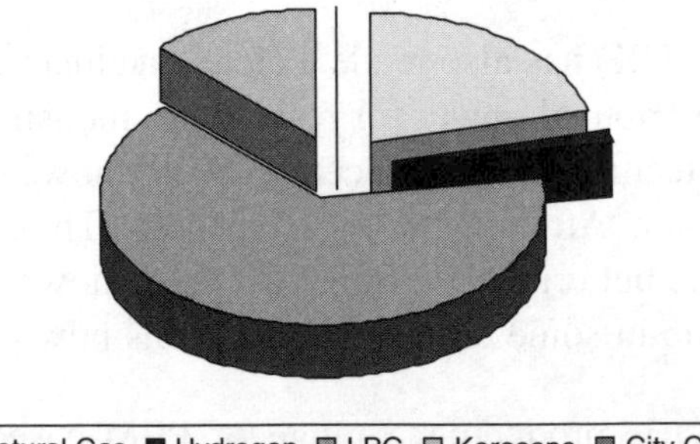

Figure 4.8 Fuel choice during 2006 for residential fuel cells [City Gas, which is mainly used in Japan, is a mix of hydrogen and natural gas, with the main component being natural gas.] *Source*: © Fuel Cell Today

which produce and source all their own power. Within this sector, fuel cells are the lynchpin providing the adopters with high-efficiency technology. This sector though, we anticipate, will not be the norm, with comparatively widespread grid parallel units being developed and deployed that can also act as primary power in blackout situations.

Case Study – H2PIA

H2PIA is a stunning architectural plan of what could be achieved using fuel cells and renewable energy. At the time of this writing, the project, based in Denmark, one of the world's most innovative and forward-looking countries in terms of fuel cell application and development, is yet to be started beyond the feasibility study. The intention is to start construction at some point in 2007.

The H2PIA concept is for an energy-independent, oil-free, hydrogen/fuel cell-based community with a mix of off-grid, grid parallel and grid-dependent housing. The three types of housing, termed 'Villa Unplugged', 'Villa Hybrid' and 'Villa Plugged', are each aimed at different adopters but all using the cleanest source of energy available.

Villa Unplugged – this represents the off-grid housing discussed earlier in the chapter where fuel cells run off hydrogen (in this example) produced from local renewable sources. Villa Unplugged will not be connected to the electricity grid (Figure 4.9).

Figure 4.9 Villa unplugged
Source: © www.H2PIA.com

Villa Hybrid – these villas are linked up to the electricity grid but draw their primary energy demand from the fuel cells, located in the houses' car. The car will receive its hydrogen from a central production plant supplied via pipelines and when stationary will be plugged into the house acting as a stationary fuel cell.

When the car is being used, the villa will revert to drawing power down from the electricity grid (Figure 4.10).

Figure 4.10 Villa hybrid
Source: © www.H2PIA.com

Villa Plugged – this represents the current paradigm in most housing developments, where the accommodation is supplied by an electricity

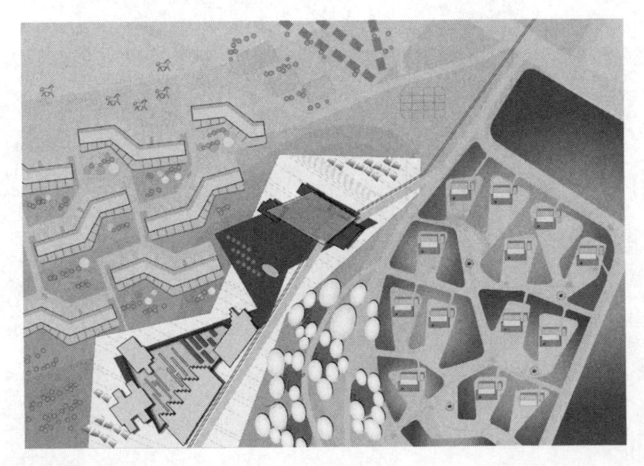

Figure 4.11 H2PIA city
Source: © www.H2PIA.com

grid from a central power plant. The difference here though is that the central power plant is also located in the development and is hydrogen based, with the hydrogen coming from local wind and solar 'energy parks' (Figure 4.11).

The project's website is http://www.H2PIA.com.

4.6 Government and zero-carbon homes

Domestic energy use contributes a significant quantity of carbon dioxide to a country's annual carbon dioxide budget. Energy consumption for heating, lighting, cooking and, increasingly, consumer electronics all contribute to a large slice of a country's carbon dioxide emission portfolio. Governments, reacting to the threat of climate change, and to issues such as energy security, are increasingly developing policy plans around zero- or near-zero-carbon homes. One example of this is the UK, where in 2006 it was announced that by 2016 all new homes must be carbon neutral. How this is to be achieved practically is yet to be worked out, but fuel cells will surely play a role.

4.7 A list of twenty suggested questions to ask yourself if you are considering using fuel cells in a residential development

General questions

1. Do you have a run down of local interconnection regulations?
2. Have you worked with the local planning authorities from the start of the project?
3. Do you have a feed-in tariff in place and can this be used in conjunction with net metering?
4. Do the local emergency services need extra training/information in case of emergencies?
5. Does the resident understand or want extra training/information? Is the resident comfortable with the project?
6. Have you considered the other alternatives?
7. Do you have access to a qualified installer?
8. Do you have access to a qualified maintenance technician?
9. Are you on good terms with the local press?
10. Do you have a lot of patience?

Fuel cell-specific questions

1. Do you want/need high-grade waste heat?
2. Do you know what size of unit best suits the development's needs?
3. Do you have a fuel cell supplier for the unit?
4. Do you prefer to use a turn-key project group?
5. Are there any local/national subsidies that may be available to install a fuel cell?
6. Do you have a good supply of a fuel which might be suitable for a fuel cell project?
7. Is there an option for renewables to be linked into the system?
8. Are you wanting to buy or lease the fuel cell?
9. Have you got a thorough breakdown of codes and standards for hydrogen storage (if using direct hydrogen)?
10. Do you feel confident that you have all the information to make the right decision for your project?

4.8 Summary

Fuel cells for use in residential applications are a lot closer to being a market reality than many, including some policy-makers, realise. Though the realisation of off-grid power is a challenging prospect, the societal and economic development benefits of this are too great to write off as simply a long-term vision. In the near term (2007–2012), we can expect to see a range of utility tie-in mCHP fuel cell-based products in circulation. Though they will more than likely be top-end market products, in terms of price, and will require clear guidance on grid connection, they could represent a steep change in thinking about home power, both for the consumer and for the energy provider.

Further reading

There are a number of easy-to-access reports that have been published since 2000 on demonstration projects, modelling and research and development (R&D) efforts. A select number of these, including those that have been of particular use in the production of this book, are listed below. This is not by any means an exclusive list.

1. Torrero, E., McClelland, R., "Evaluation of the Field Performance of Residential Fuel Cells", National Renewable Energy Laboratory (NREL), Report Number NREL/SR-560-36229, 2004, downloadable from the US Department of Energy Information Bridge

2. Braun, R.J., Klein, S.A., Reindl, D.T., "Assessment of Solid Oxide Fuel Cells in Building Applications. Phase 1; Modelling and Preliminary Analyses", University of Wisconsin-Madison, 2001
3. Tiax, "Grid-Independent, Residential Fuel-Cell Conceptual Design and Cost Estimate", Report Reference: 76570, 2002. Full report downloaded at "http://www.netl.doe.gov/technologies/coalpower/fuelcells/publications/gridindepentresident
4. David, M.W., Fanney, A.H., LaBarre, M.J., Henderson, K.R., Dougherty, B.P., "Parameters Affecting the Performance of a Residential-Scale Stationary Fuel Cell System", National Institute of Standards and Technology, Building and Fire Research Laboratory, 2005

5 Uninterruptible power supply/backup

5.1 Introduction

This chapter aims to look at the use of fuel cell technology for uninterruptible power supply (UPS) services. Interestingly, this application is already one of the first adopters of fuel cell technology.[1]

In general, UPS systems provide security against the loss of power in a grid failure, or grid fluctuations, either sags or spikes.[2] In the case of power loss, the UPS system is scaled to provide one of two functions – either enough power to safely shut all systems down once work has been saved or extended run-time UPS systems that provide continuous power during the period in question.

Two further system classifications are off-line and on-line systems. An off-line UPS is one that it is off-line all the time the utility is providing power and which kicks in, instantaneously, when this fails. On-line systems conversely provide power all the time, with the grid taking the strain if the power generator fails.

As discussed in Chapter 4, as companies are becoming increasingly information intensive and the costs, and wider impacts, associated with downtime, more widely understood UPS systems are diffusing more rapidly into the

[1] One small note to make here is that this text does not cover any portable fuel cells that are being used for backup power. Portable fuel cells are defined, for the purposes of this book, as any unit that is specifically designed and packaged to be moved, independent of its size or output range.

[2] A sag is a transient drop in electricity voltage and a spike is a transient over-voltage. Whereas a sag can cause a flickering of electrical equipment, a spike can cause damage.

marketplace. This increased diffusion is also having the impact of bringing adoption costs down, further enhancing the rate of uptake.

5.2 Development to date

Figure 5.1 shows the size of the UPS market as a share of the total fuel cell industry to date.

What can be seen from this figure is that this specific application is growing steadily, with last year seeing a nearly 90% market increase. This is highly indicative of the level of application 'fit-for-purpose', i.e. a technology providing a solution to a current market need. Unlike residential and larger distributed generation, there is little or no governmental backing or support for this application. The interest is coming directly from the industry, which is looking for a technological option that provides the characteristics that they require at a price point they are willing to pay.

5.3 Incumbent technologies

As with nearly all other applications for fuel cells, the UPS sector already has incumbent technologies with which it is competing for market space. For the sake of clarity, a brief run-through of the two main incumbents, batteries and gensets, is provided.

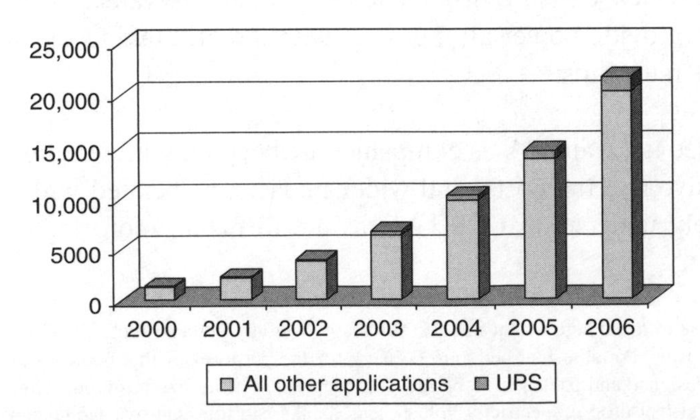

Figure 5.1 UPS market size to date as compared with the cumulative fuel cell industry

5.3.1 Batteries

Sealed valve-regulated lead acid (VRLA) batteries are often used in operations in enclosed spaces where the emissions from diesel gensets are unacceptable. Battery rooms, which are literally rooms full of batteries for UPS, is a common site in buildings such as data centres and telecommunications facilities. The sealed batteries are scaled by linking them together in series, known as a 'string', with the number depending on the amount of power required and size of the batteries.

One of the key disadvantages with batteries is their sensitivity to temperature. According to the British firm UPS systems, VRLA batteries are designed to operate at an optimal temperature of 25°C, with every 5°C above that equating to a halving in battery life. Batteries normally have a 5- or 10-year lifetime with replacement due a year or so before it is out of life. They are also designed to be 'hot swappable', which means that one unit can be taken out of the series whilst it is running and be replaced without system performance degradation.

Economics of battery UPS

As with any power installation, cost varies according to a number of site-specific factors. List prices for VRLA batteries tend to be in the region of US$340 per kWh, with turn-key project prices being substantially higher.

Advantages and disadvantages of battery UPS systems

The main advantages of using battery technology for UPS are:

Zero emissions
Understood costs
Low maintenance
High reliability
Easy to scale

The main disadvantages of using battery technology for UPS are:

Size
Limited operating time
Potentially high capital costs depending on run-time required

5.3.2 Gensets

Generator sets are internal combustion-based units that run off diesel, natural gas or propane. As with any engine, the moving parts in the units cause wear and tear and, to a certain extent, gensets have to be heavily maintained. Also, as with any combustion of fossil fuels, gensets produce harmful emissions that can cause issues in enclosed spaces. They are a mature, understood technology and can be bought at a low capital cost. Compared directly with batteries, gensets also tend to have a longer run-time as it is easier to increase the quantity of fuel stored for a genset than increase the number of batteries in series.

Emissions

One of the often-cited disadvantages for using fossil fuel gensets is the emissions caused by using the units. Increasingly, legislation in countries such as the USA is tightening the legal acceptable limits of pollution from stationary sources. By the time this book is published, stationary compressed ignition engines manufactured in the USA will have to be compliant with Tier 1 emissions from non-road engines. Cutting through all the jargon, what this means is that new gensets cannot produce more than:

- 8 g/kWh CO, 10.5 g/kWh NMHC + NOx, 1 g/kWh PM for engines <8kW
- 6.6 g/kWh CO, 9.5 g/kWh NMHC + NOx, 0.8 g/kWh PM for engines 8 – <19kW
- 5.5 g/kWh CO, 9.5 g/kWh NMHC + NOx, 0.8 g/kWh PM for engines 19 – <37kW
- 9.2 g/kWh NOx for engines 37 – <130kW
- 11.4 g/kWh CO, 1.3 g/kWh CO, 9.2 g/kWh NOx, 0.54 g/kWh PM ≥130 kW

These are US Federal standards, with some states within the US having stricter targets. In a number of other countries, non-portable or non-road diesel-based internal combustion engines are simply not regulated.

Economics of gensets

Though costs for specific units vary, the industry standard for capital expenditure is around US$200–300 kW, although costs of up to US$400 kW have been quoted in the literature.

Other costs that need to be taken into account are naturally dependent on a number of factors such as:

- Installation costs
- Engine efficiency and time operated – quantity of fuel needed
- Delivered price of fuel
- Regular maintenance schedule
- Lifetime of the equipment
- Whether any of the waste heat can be used to offset space heating/cooling bills

Total cost of ownership (TCO)[3] for a diesel genset is predominantly delivered fuel and onsite maintenance costs.

Advantages and disadvantages of generators

Advantages of using gensets include:

Reliable, if maintained properly
Long life of the capital equipment
Longer run-time than batteries
Low capital costs
Mature technology

Disadvantages to using gensets primarily are:

High maintenance
Noisy
Emissions (quantity depends on fuel used)
Require fuel storage
Start-up time

5.4 Fuel cells for UPS systems

During 2006, over 1000 fuel cell UPS systems were ordered or delivered across the globe. To put this in context, this represents nearly 10% of all

[3] TCO covers engineering costs, capital costs, installation and start-up costs, maintenance costs, fuel costs, saved energy and can also include charges from the utility for stranding.

units during 2006. Countries such as Germany, South Africa, the UK, the USA and the continent of Antarctica are now employing fuel cell systems in a UPS capacity. The reason is that the technology displays most of the benefits of using batteries or gensets, such as low maintenance, reliable and easy scalability, without the disadvantages such as emissions, noise, footprint or temperature issues.

Currently, three types of fuel cells are being used or considered for UPS systems. For an explanation of the technology underneath these types, I refer you back to Chapter 2.

5.4.1 Proton-exchange membrane (PEM)

To date, this type of technology is used in over 99% of new UPS units. Not only is PEM technology close to being commercial in this sector in terms of cost, manufacturing capability and distribution agreements, but it also experiences synergistic benefits from other fuel cell applications in terms of development of codes and standards, press coverage and growing consumer awareness.

Developers and distributors

There are a growing number of companies that are producing or distributing PEM fuel cell UPS units. Four of the biggest players and their products are profiled below:

1. APC (Denmark), which uses Hydrogenics stacks, has launched a fuel cell backup power system for data centres. The 'InfraStruXure', with rack mounted integrated fuel cell system, provides up to 3×10 kW of power. The hydrogen fuel is bottled and stored in T-shaped gas cylinders, with each cylinder storing enough fuel for 79 minutes per 10 kW stack. The system is available now from a range of distributors such as UPS Systems (UK).
2. Rittal (Germany) uses IdaTech stacks in its unit which is designed for outdoor use and another for data centres and indoor use. The initial offering was a 3 or 5 kW rated power unit and then, in 2006, a 30 kW, scalable to 60 kW was released.
3. Plug Power is one the best-known fuel cell manufacturers and also one of the few companies developing a business model that sees the use of its own stacks fully integrated into the end-use product. Plug Power units are being distributed globally by a number of companies, as well as by

Plug itself in North America. Distributors include IST (South Africa), Logan Energy (North America and Europe) and siGen (UK). The main product line for UPS systems is the GenCore product line and includes dedicated units for outdoor and indoor use, remote telecommunications sites to data centres.

4. ReliOn has undertaken substantial testing of its units in varying environmental conditions (see *Case Study* later in this chapter for an example) (Table 5.1).

Cost (current and target)

As can be seen from Table 5.1, current price list costs are higher than both the VRLA and genset options. A number of studies have been published on the TCO of the three options, with the output results varying from study to study, one showing a positive case for fuel cells and the next not. What this means is that there are currently no hard-and-fast rules on costs, with different systems from different manufacturers having different cost structures.

Example of studies that list costs are:

1. '*Alternative Power Generation Technologies for Data Centres and Network Rooms*', by APC

Using a number of assumptions, APC modelled the TCO for generators, fuel cells and another up and coming technology, microturbines. The modelled costs are shown in Figure 5.2.

What this figure apparently shows is that under all conditions, fuel cells come out most expensive compared to rival technologies.

Note: When the data from this model are unpacked, it is clear that the costs for the fuel cells are dominated by the capital cost, modelled here are more than twice that of the generator. Cost targets aim at bringing the per kilowatt costs down to under US$200, which, if reported data are to be believed, developers are not far from achieving.

2. '*Fuel Cells in Backup Power Applications*', Federal Energy Management Programme

This study compared the economics of using a standard VRLA unit and an off-the-shelf 1 kW PEM fuel cell unit. Both sets of results are summarised in Table 5.2.

Table 5.1 Comparison of technical specifications

	APC ("Infra-StruXure")	Rittal (fuel cell UPS)	ReliOn ("I-1000")	Plug Power ["GenCore" (model 5B48)]
Fuel cell				
Power out (kW)	Up to 30	Up to 5	1, scalable	Up to 5
Rated voltage (V)	200	230	24 or 48	48
Voltage output (V DC)		46–54		42–60
Fuel type				
Fuel	Hydrogen	Hydrogen (or natural gas or methanol, with a reformer)	Hydrogen	Hydrogen
Fuel purity		99.999	99.95	99.95
Consumption (standard liters per minute)		75 15 @ 1 kW	7.7 @ 0.5 kW 75 @ 5 kW	40 @ 3 kW
Hot swappable	Yes	Yes	Yes	Yes
Enclosure				
Dimensions (mm)	2070 × 597 × 1092	650 × 1345 × 650	44.5 × 69 × 51	112 × 66 × 61
Weight (kg)	400	182	66	227
Noise (dB(A))		60	53	60
Operating conditions				
Temperature range (°C)	+5 to +40	−40 to +50	0 to +46	−40 to +46
Relative humidity (%)	0–95	0–95, no condensate	0–90	0–95, no condensate
Height above sea level (meters)	0–9	0–2	−60 to +4000	−60 to 1829
List price (2006)	40,000 for a 10 kW system		8050 per 1 kW system (2004)	15,000 for the 5 kW system
Installed price (2006)			31,000	26,000

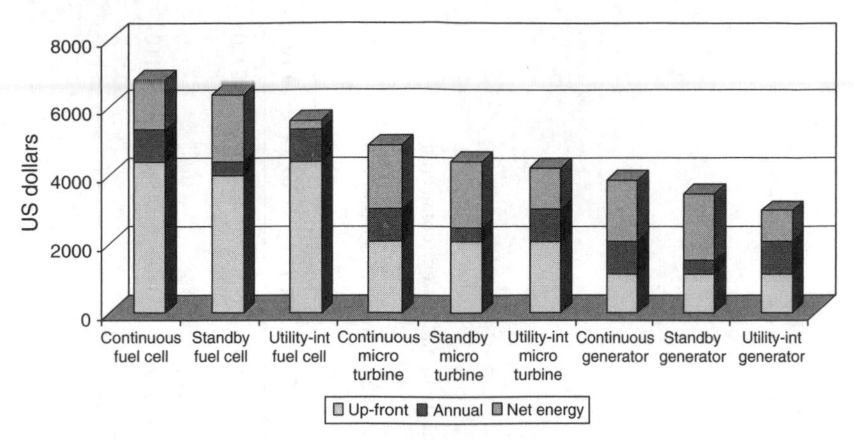

Figure 5.2 Total cost of ownership as modelled by APC
Source: © APC

Table 5.2 Comparison between backup power systems with 48-hour capacity

Backup power system	Initial system cost ($)	Maintenance cost ($)	Battery replacement cost ($)[b]	PV lifecycle cost ($)[c]
VRLA battery	29,000	600/year + 300/2 years[a]	20,260	53,250
PEM fuel cell	31,000	600	1100	36,457

VRLA, valve-regulated lead acid.
[a] Performed capacity test every other year.
[b] 5-year system life expectancy for VRLA batteries.
[c] Based on 10-year life expectancy.

One thing to note from Table 5.2 is that the 'initial system cost' is a turn-key cost, and therefore includes all installation costs.

3. "*Proton Exchange Membrane Fuel Cell Demonstration of Domestically Produced PEM Fuel Cells in Military Facilities*", US Army Core of Engineers, Engineer Research and Development Centre

This report analysed four units in the field, as compared with VRLA batteries in terms of both performance and economics. The results, in this study, showed a saving of 27% using the fuel cell technology (Table 5.3).

The explanation for the discrepancies between the different studies is the difference in the assumptions behind them.

Table 5.3 Economic results from US army testing of valve-regulated lead acid (VRLA) battery set and PEM fuel cell unit

System requirements:
Load: 4 kW
Run-time: 12 hours maximum
Discount rate: 6%

Option A: Engine generator with battery bank
Cost assumptions

Engine Generator	$10,120 (based on Cummins Onan GCAC-1385 generator)
Automatic transfer switch	$2,400 (industry data)
Fuel storage tank and monitor	$1,518 (15% cost of generator system)
Battery stack	$6,720 (based on $0.35/watt-hour; 5-year replacement)
Installation	$20,400 (based on fuel spill containment, generator install, electrical tie in and battery connection)
Fuel service	$650 (based on monthly fuel cycling)
Annual maintenance	$3,740 (genset = $225/month per MA/com, Battery = quarterly service of 4 hours @ $65/hour)

Cash flows	**Y0**	**Y1**	**Y2**	**Y3**	**Y4**	**Y5**	**Y6**	**Y7**	**Y8**	**Y9**	**Y10**
Capital equipment ($)	41,808					8720					8720
Other initial expenses ($)											
Annual expenses ($)		4390	4390	4390	4390	4390	4390	4390	4390	4390	4390
Net cash	(41,808)	(4390)	(4390)	(4390)	(4390)	(13,110)	(4390)	(4390)	(4390)	(4390)	(13,110)
NPV	(85,504)										

Option B: 4 × 1 kW PEM fuel cells

Cost assumptions

4 × 1 kW PEM fuel cells	$32,200 (ReliOn, based and list price)
1-hour battery Bridge	$1,680 (based on $0.35/watt-hours; 5-year replacement)
ReliOn outdoor enclosure	$7,500 (list price)
Replacement cartridge	$300 (6-year replacement 1 × $300)
Installation	$9,000 (includes site preparation, pad and conduit)
Annual maintenance	$520 (fuel cell system and batteries, two visits per year @ 4 hours/visit = $520)
Annual hydrogen fuel cost	$585 (six cylinders; $5 per bottle per month, fuel and delivery)

Cash flows	**Y0**	**Y1**	**Y2**	**Y3**	**Y4**	**Y5**	**Y6**	**Y7**	**Y8**	**Y9**	**Y10**
Capital equipment ($)	50,965					2200					2200
Other initial expenses ($)											
Annual expenses ($)		1105	1105	1405	1105	3305	1405	1105	1105	1405	1105
Net cash	(50,965)	(1105)	(1105)	(1405)	(1105)	(5505)	(1405)	(1105)	(1105)	(1405)	(3305)
NPV	(62,611)										

Fuel cell saving 27%.

All of these costs are case-by-case and therefore potential adopters have to work the economics on an individual approach.

5.4.2 Solid oxide fuel cells (SOFCs)

SOFCs are a high-potential technology for standby power. SOFCs are high-temperature fuel cells which are experiencing intense levels of research and development (R&D) which in turn is having the impact of bringing unit sizes and costs down, whilst increasing power densities. The US-based SECA programme is instrumental in laying out targets and timelines for these developments.[4]

Benefits of using SOFC technology

Fuel flexible. SOFC units can use 'dirty' hydrogen that has been reformed from fossil fuels

Potential disadvantages of using SOFC technology in backup power applications

Longer start-up time due to high-temperature technology
High temperature off-gas which would need to be actively managed

Among present companies, the main player working on this option is Acumentrics.

5.4.3 Alkaline fuel cells (AFCs)

Though AFCs are on the fringe of the fuel cell industry in terms of units developed and delivered, a few companies are still pursuing the technology and developing UPS applications.

---

[4] The solid energy conversion alliance has targets of:

- Reductions in cost to US$400/kW by 2010 for 3–10 kW stack
- Efficiency of 40–60% (based on Lower Heating Value)
- A minimum of 50,000 units by 2010
- The development of a modular, solid-state fuel cells that can be mass produced for different uses
- Availability over 80%

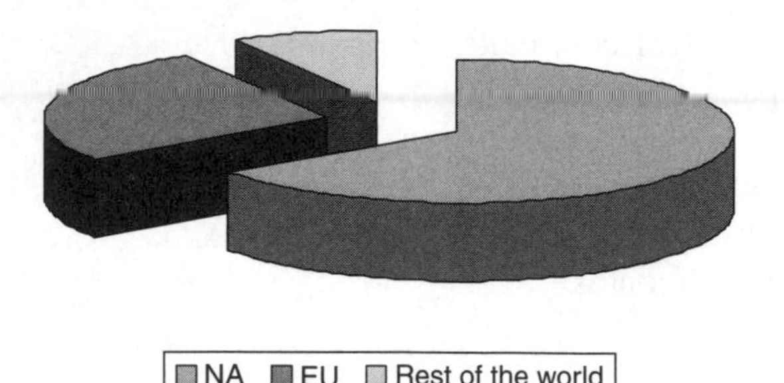

Figure 5.3 2006 fuel cell uninterruptible power supply (UPS) systems by region of use
Source: © Fuel Cell Today

5.4.4 Molten carbonate fuel cells

Molten carbonate technology is another high-temperature fuel cell. As it is being looked at as primary power using the grid for backup, it is covered later in the book in Chapter 6.

5.5 Geography

Different regions of the world have varying levels of interest in UPS systems. In Japan, were the electricity grid is not only ubiquitous but also very stable, interest is higher in systems that take some of the primary load off the grid, e.g. residential systems, rather than providing power in the case of grid failure. In Europe and North America, where the grid is also widespread but not as stable, interest in UPS is intense. The third level of interest comes from those regions where either there is simply no grid electricity or it is very unreliable. South Africa is an example here where though there is a grid, fluctuations often occur (Figure 5.3).

Case study – Using PEM fuel cells as backup power for telecommunication sites in Italy

To date there have been a number of projects and studies, which have used fuel cells either to validate data or to run commercially. One such study was undertaken in Italy, eight of ReliOn I-1000 units. The short case study below outlines the main inputs and outcomes from the project.

Basics:

Fuel cell manufacturer: ReliOn
Distributing agent: SGS

Contractor: Telecom Italia
Fuel cell type: PEM unit
Fuel: Gaseous hydrogen
Configuration: Scalable 1 kW
Fuel cell cost: US$8050 kW (2004 list price)
Test lengths: months

Background:

During 2005, Telecom Italia contracted the Italian distributor agent of fuel cell manufacturer ReliOn for a number of its I-1000 units to test reliability as backup power for telecommunication sites. Eight I-1000 1 kW units were installed at three Telecom Italia plants in the configuration of four units (4 kW) at one site and two further sites of two units each.

The main pieces of hardware at the sites were the fuel cell, and a small battery array, which were housed in indoor enclosures with external venting of the heat produced. The hydrogen was stored externally in high-pressure storage vessels, each containing 108 Nm^3 H_2 at standard temperature and pressure (STP). In addition to these, standard safety equipment was also installed.

Test phases

The study was set up to go through four test phases, each looking at critical metrics such as start-up time, time to full load and overall reliability. The four phases after installation, initial testing of safety systems, calibration and testing of the remote transmission of data were:

Phase 1: Starting and stopping of the systems at random times for differing durations over a period of approximately 100 times to verify system behaviour. This phase ended with the emptying of the initial hydrogen canisters.

Phase 2: Testing of the long-run capabilities. This specifically included data to calculate system efficiency

Phase 3: Standardised backup power tests, including:

- Attainment of nominal conditions of the system
- Fast turn on/off cycles
- Recharge to float voltage after heavy discharge of the battery packs

- Repeated and frequent grid disconnection
- Automatic start-up procedure in case of low battery voltage
- Start-up of the units after long inactivity

Phase 4: Replacement of the existing battery packs with smaller ones to test transient start-up.

Results of tests and project outcomes:

- The base conclusion from the study was that under all conditions tested, the fuel cells operated with full reliability.
- The measured electric efficiency of the units was between 38.4 and 39.1%
- The fuel cells required, as predicted, minimal maintenance
- The units were stated to be an economically viable choice

The short-term outcome of the project was that Telecom Italia expanded the project to include a further 20 sites.

The full project report can be read at:

Tomais, A., Concina, M., Grossoni, M., Caracino, P., Blanchard, J., *"Field Applications: Fuel Cells as Backup Power for Italian Telecommunication Sites"*, presented at Intelec 06, 10–14th September 2006, Rhode Island, USA.

5.6 A list of twenty suggested questions to ask if you are considering using fuel cells as in a UPS application

General questions

1. Do you have a run down of local interconnection regulations?
2. Do you have any site-specific issues that rule out, or rule in, fuel cells?
3. Are you only looking to level spikes and sags?
4. Are you looking for enough power to shut down equipment without losing work or are you looking for full cover during the grid failure?
5. How long do you want in terms of emergency time covered?
6. Does the local emergency services need extra training/information in case of emergencies?

7. Have you considered the other alternatives?
8. Do you have access to a qualified installer?
9. Do you have access to a qualified maintenance technician?
10. Do you have a lot of patience?

Fuel cell-specific questions

1. Will the production of waste heat cause a problem?
2. Do you know what size unit best suits the development's needs?
3. Do you have a fuel cell supplier for the unit?
4. Do you prefer to use a turn-key project group?
5. Do you have a regular supply of hydrogen or methanol?
6. Is there an option for renewables to be linked into the system?
7. Are you wanting to buy or lease the fuel cell?
8. Have you got a thorough breakdown of codes and standards for hydrogen storage?
9. Do you feel confident that you have all the information to make the right decision for your project?
10. Is there any site-specific circumstances that could rule in/out fuel cells?

5.7 Summary

Fuel cell technology is entering the UPS application sphere because it offers a number of advantages over current systems. At present, the technology is being used when these advantages overcome the main contemporary disadvantage of high capital costs.

Though not the only new technology on the horizon for this application, fuel cell usage is growing fast in the UPS sector with the expectation that it will continue to grow as manufacturing is ramped up, costs come down and the fuel supply chains materialise.

At present, the only real UPS market option is the PEM fuel cell. The vast majority of manufacturers in this space are working on PEM units and have built up solid initial distribution networks for their products. Interest from the US military is helping to push developments in PEM technology but also SOFC, which should have the impact of seeing SOFC UPS systems in the market in the not-too-distant future.

If the predicted increase in electricity grid instability continues in areas such as North America (USA and Canada) and Europe, as well as the growing industrialisation of nations such as South Africa occurs then the market for UPS systems is also predicted to increase in line with this. With the interest in fuel cells for this application already high, we can expect to see system numbers also increase.

6
Commercial buildings and power plant fuel cells

6.1 Introduction

This third type of stationary fuel cell application is split into two main sections: (1) power to non-residential, commercial, buildings to provide electricity, and in many cases heat. This can be anything from an office block or convenience store and (2) distributed generation (DG) power plants providing electricity for the electricity distribution network.

These applications tend to be grouped together due to the size of the units at >10 kW and up to multi-megawatt (MW) capacity. Also due to this very large scope for adoption, there are units currently being used in a wide variety of settings and sizes, from 10 kW right up a, proposed 5 MW and beyond. Often though, these larger (>1 MWe) sizes are base modules of 200, or 250 kW, units linked in series.

This chapter differentiates the fuel cells into two main business segments: power to buildings, such as office blocks and hospitals, and pure utility-generating capacity. As with the other chapters, this chapter looks at the technology under development, its medium-term outlook, the main companies presently involved and the current and medium-term economic situation.

6.2 Development to date

Of all the three major applications discussed in this book, this is the application that has the longest history of adoption, with UTC's phosphoric acid fuel cells (PAFCs) being bought and sold back in the early 1990s in a major (though temporary) buy in programme by the US military. Figure 6.1 shows

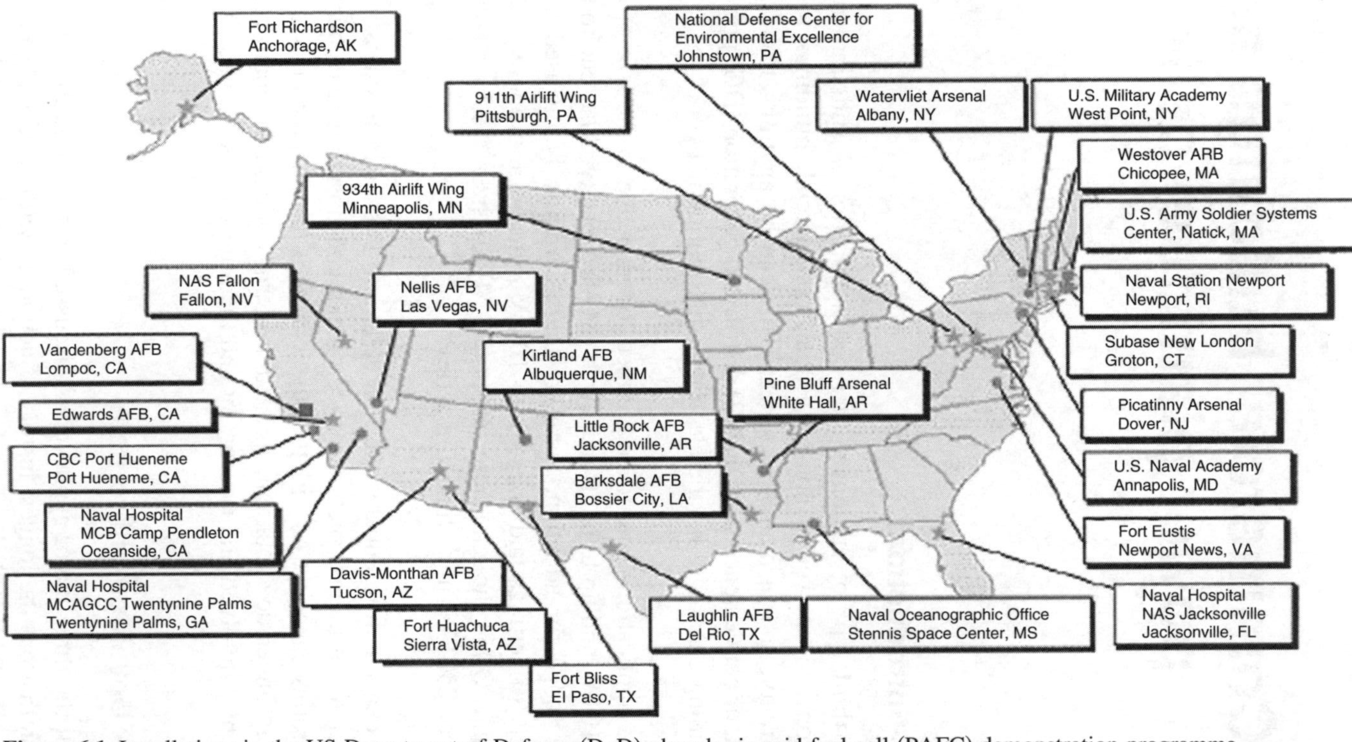

Figure 6.1 Installations in the US Department of Defense (DoD) phosphoric acid fuel cell (PAFC) demonstration programme
Source: © US DoD

a map of Department of Defense (DoD) PAFC installations during FY93 and FY94.

During the late 1990s, interest in the application tailed off, probably due to a lack of complementary drivers and the perceived slow rate of technological progress that PAFC were experiencing at the time.

What this early interest did was to create a pool of knowledge that still exists for large fuel cell project installation, operation and maintenance. As with many other fuel cell applications in the USA, the majority of this experience rests within the US Military, but a growing number of consultants and fuel cell turn-key operators have appeared offering sales, installation, operation and maintenance services.

Figure 6.2 reiterates graphically that at an early stage in the development of fuel cell markets, large stationary units had a significant percentage of the total units adopted, but this has over time decreased slightly. As interest switched focus in the 1990s to transport, primarily, a number of alternative fuel cell formats started to dominate.

Since 2000, there has been something of a balancing-of-application interest with a number of technological breakthroughs being made and subsidies for adoption now available.

In terms of numbers, the last 6 years (since 2000) has seen well over 300 new units being added to the pool, with the average size jumping from the

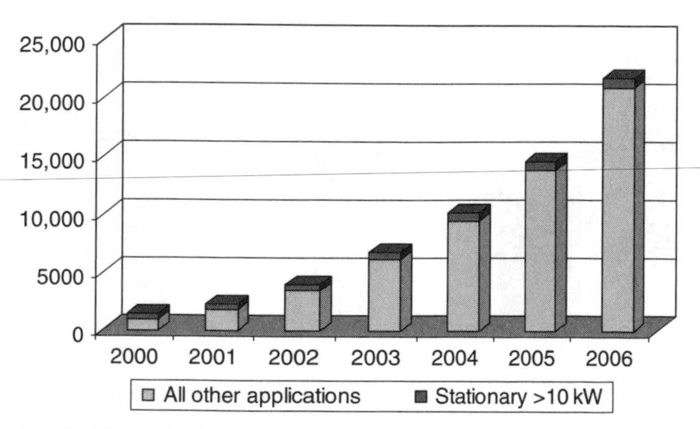

Figure 6.2 Fuel cells >10 kW as a share of the total fuel cell cumulative pool
Source: © Fuel Cell Today

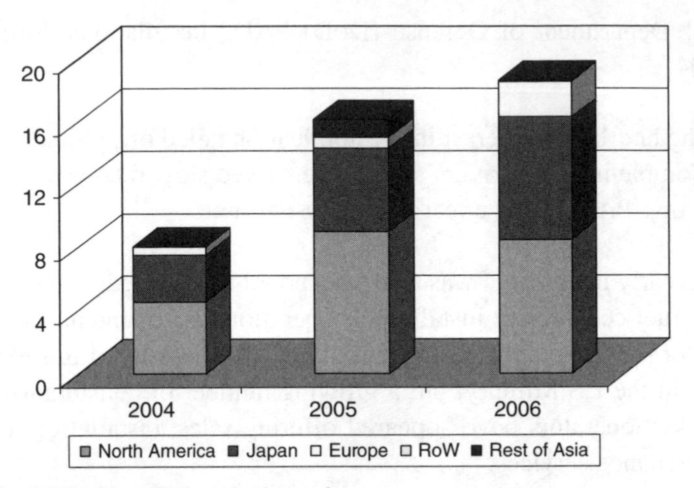

Figure 6.3 MW of installed power by region
Source: © Fuel Cell Today

low tens of kilowatts to approaching a third of a MW today. With a growing number of multi-MW projects in the pre-financing stage, this trend of size per unit/project increase can be expected to continue (Figure 6.3).

If we turn our attention to where these units are located geographically, Figure 6.4 highlights that at present North America (the USA and Canada combined) and Japan are the two key markets.

Looking forward, key governmental policy drivers, which are discussed in detail in Chapter 7, show that in the coming decade South Korea and the European Union are likely to see significantly higher rates of adoption

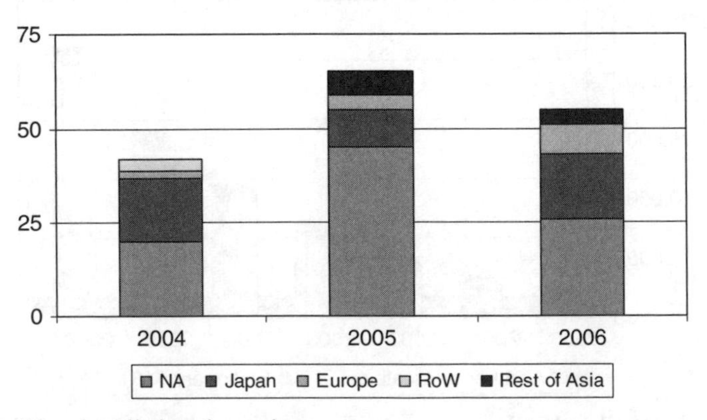

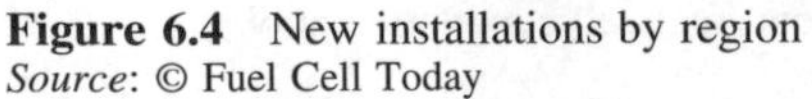
Figure 6.4 New installations by region
Source: © Fuel Cell Today

than at present. In summary, South Korea has a government policy of 300 stationary fuel cell power plants by 2012 and within Europe the 7th Framework Programme (FP7) has an outline for the adoption of some 2650 units by 2015, 60 of which are to be larger than 1 MW in capacity. Again to put this into context, over the last 10 years some 100+ MW has been adopted worldwide. If the plans in South Korea and Europe alone succeed, we would see upwards of 400 MW[1] new installed capacity from stationary fuel cells.

One other development trend of note at the post-1 MW scale is the development of modular fuel cells linked in parallel to a gas turbine. The purported benefits of this form of hybridisation include higher efficiency, lower life-time costs and higher recovery of waste heat.

6.3 Technology development

Of any fuel cell application to date, this is the area with the largest number of fuel cell types being developed. In other words, molten carbonate fuel cell (MCFC), PAFC, proton-exchange membrane (PEM) and solid oxide fuel cell (SOFC) units are all being developed and tested for adoption in scales of >10 kW. The only mainstream fuel cell that is not being developed for these applications is direct methanol fuel cell (DMFC).

Note that Figure 6.5 represents units installed and not MW installed.

What can be seen from the figure is that currently the field is wide open for the four technologies listed above. It is impossible to say at this stage of market development if any single technology will become dominant enough to lock-out the others, but it is likely that if this does occur it will not happen for a decade or so to come. The reason for this is that the companies behind the technologies are working hard on R&D to bring down costs, build manufacturing capability and increase adopter confidence and are working together on codes and standards. Also each of the technologies is finding separate areas within the market in which it has an optimum technological fit.

[1] This 400 MW is a very conservative estimate based on all units being no greater in size than 10 kW apart from the stated 60 at larger than 1 MW.

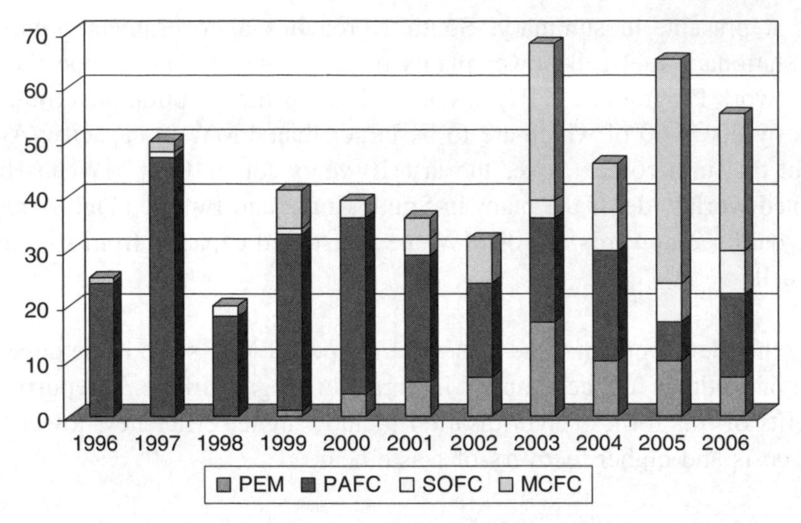

Figure 6.5 Technology adoption to date of units >10 kW
Source: © Fuel Cell Today

An example of this is using large PEM fuel cells (PEMFCs) in chemical plants. One such project is a collaboration between Nedstack's, a fuel cell manufacturer, fuel cell technology and an Akzo Nobel chemical plant. The plant produces as products chlorine (Cl_2) and sodium hydroxide (NaOH) and as a waste stream high-purity hydrogen. During the project, Nedstack will put in place a 120-kW peak PEM fuel cell pilot power plant,[2] which by utilising the hydrogen will produce electricity which is then fed back directly into the plant. The pilot plant described above is of a modular design allowing a later scaling upto MW size.

This pilot PEM unit will demonstrate that via regeneration of hydrogen, 20% electricity consumption by the chemical plant can be diminished. The demonstration project is being subsidised by the Dutch Ministry of Economic Affairs, within the EOS programme. The today's (2006) cost of the PEMFC power plant is calculated to be €1000/kW. NedStack expects that the PEMFC power plant will generate competitive electricity compared to the grid, at 2009/2010. This cost competitiveness can be obtained if market demand meets increases as projected, allowing a further ramping up of production-inducing scale economies to being cost downs.

[2] To date, a fuel cell power "module" is in place.

6.4 Co-generation for early market such as hospitals, hotels and swimming pools

What is probably surprising for many who think that fuel cells are only for niche applications is that to date a large number of units are currently powering hotels, swimming pools and hospitals – all early, high-value, markets. Working on the same principle as micro-combined heat and power (mCHP) residential fuel cells, providing electricity and heat to a specific building, or buildings, these units are becoming ever more popular as the economics and security of supply, in terms of reliability and market-based drivers, become more attractive. Figure 6.6 illustrates this in a representation from one of the key market players, FuelCell Energy, of the breakdown of its units by end user.

6.4.1 Hospitals

Hospitals are a fairly unique type of building in that they not only require a constant, non-interruptible, supply of electricity and heat but that the heat can be in the form of space heating or superheated steam for equipment sterilisation. Three companies, FuelCell Energy, MTU CFC and UTC are already active in this area with MTU CFC providing units with a high-temperature steam outlet which can be fed directly into a hospital's steam distribution network.

To date, around a dozen hospitals have either used fuel cells in the past or are still using them.

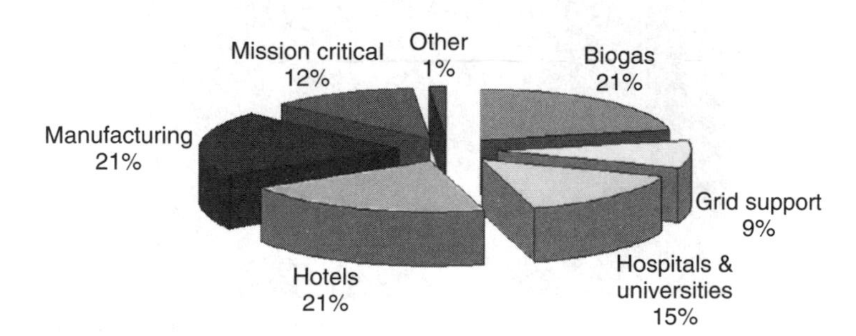

Figure 6.6 FuelCell Energy deployed units end-user markets
Source: © Fuel Cell Energy

6.4.2 Hotels

Like hospitals, hotels require a constant supply of electricity and heat. Unlike hospitals though, there are also a large number of hotels that operate in chains. Hotels such as the Sheraton and Westin (both under the Starwoods Hotel and Resorts management family) already have installed fuel cells in a number of locations, primarily in Connecticut and New York (Figure 6.7).

6.4.3 Swimming pools

Woking Borough Council, in the UK, made headlines back in 2002 by its use of a 200 kW PAFC, providing electricity to local private wires and using the low- and high-grade waste heat to help heat a local swimming pool. Since then, a number of such units have been installed globally utilising the waste heat as a community resource.

These three examples, hospitals, hotels and swimming pools, are all examples of *high-value early adopter markets* where the particular benefits that

Figure 6.7 FuelCell Energy molten carbonate fuel cell (MCFC) installation at the Sheraton Edison
Source: © Fuel Cell Energy

a fuel cell can bring outweigh the initial barriers in terms of upfront cost and increased paperwork. As manufacturing capability, stack reliability and adopter familiarity increases, we can expect to see these barriers reduce and more CHP adopters come forward.

6.5 Fuel cell grid-supporting power plants

Fuel cell technology represents decentralised energy-generation capability. They are modular in design, can use a variety of fuel feedstocks and in theory operate around the clock with minimal maintenance. It is a logical step, therefore, that some utilities are adopting fuel cells as decentralised power plants alongside more traditional options such as coal-fired, nuclear power and natural gas-fired plants. The basic principle of grid-connected power plant units is that of a number of fuel cell modules connected in series to each other and to the grid.

The largest fuel cell plant planned, to date, is being designed by GE Energy as a 100 MW SOFC with coal as the fuel source. In between that and the 10 kW units being used, there is a whole range of sizes for power generation being developed.

One of the major drivers for this type of decentralised power generation is the modular capability of fuel cells. Using a traditional centralised power station model, the addition of new generating capacity causes an unbalancing of supply and demand.

As the growth in demand is fairly smooth but the addition of, say, a nuclear plant can add a further 1000 MW to the grid, there will be an over-supply of electricity until demand catches up with this new installed capacity. When lead time for a new plant is also taken into account, which can be in excess of 10 years for nuclear power plants, this creates a complex system to manage. Fuel cells on the other hand can be installed comparatively rapidly, typically less than a year from order, and in smaller modular increment, of 200 kW upwards, allowing closer load following. Also critically in favour of fuel cells is the very short decommissioning time, again using a modular approach.

Another major DG driver (potentially FC DG) is grid reinforcement deferral. Adding DG to a grid network can mean that distributed network operators (DNOs) do not have to invest in highly expensive new low/medium-voltage wires when new capacity is brought online, as the DG source located close to the user reduces the net flow through the system. There are difficulties with this however: under many reward systems set up in liberalised markets around the world, distribution companies are financially rewarded for maintaining the stability of their networks. Installing DG in such a network, arguably, destabilises it. This is the nature of the trade-off between the two opposing pressures of grid reinforcement deferral and system stability.

This leads on to a potential second barrier to adoption for FC DG which does remain and could potentially be a genuine show-stopper: in some regions some utilities are prohibited from owning their own generating capacity. Where this happens fuel cells, with their potential to decrease the need for grid electricity, are a competing technology for their business and like any business it must be assumed that they will respond accordingly.

As distributed generation is an increasing trend, and one which is expected to continue to strongly develop, adoption of fuel cell power plants of the multi-MW size will also increase over the next decade. The regulatory barriers above must be addressed though before the full potential of this generating model can be realised.

6.6 Economics

Of the three areas discussed in this book, this is the one area where the current economics are best understood with the clearest short-term developmental targets.

6.6.1 Current economics

Currently, there are three main products in the open market place. These range from a modular 100 kW PEM stack to a post-1 MW MCFC unit.[3] Table 6.1 outlines the current (2006) costs of these units, which as can be seen range up to 4200 $/kW.

[3] Remembering that fuel cell units, of the same electrolyte, can be linked in series this creates a design flexibility so that the product can be tailored in a bespoke manner to the size, fuel use and emission output required by the adopter.

Table 6.1 Current prices of non-residential stationary fuel cells

Company	Unit name	Technology	Size (kW)	Price 2006 (US$)
UTC	PureCell	PAFC	200	4200
Siemens		SOFC Hybrid		In development
Rolls Royce		SOFC Hybrid		In development
Hydrogen LLC		PAFC	400	In development
Nedstack		PEM	100	(waiting for cost confirmation)
FuelCell Energy	DFC3000	MCFC	250	3200–3500

MCFC, molten carbonate fuel cell; PAFC, phosphoric acid fuel cell; SOFC, solid oxide fuel cell.

Note that these costs are based on actual project costs and not theoretical modelled costs.

6.6.2 Future economics

As with the other application areas discussed in this book, the developers in question have a range of cost targets that they feel they can hit upon. The difference with this application is that many of the developers are talking about a date of 2008 for full commercial activity. This really is only a very short time away. For the developers to realistically hit this date, they will have to do two things:

1. Increase manufacturing capability to a level where orders can be fulfilled within an acceptable time period – to date only two companies, FuelCell Energy and UTC Fuel Cells, have anything like the sufficient level of manufacturing capability. A number of other companies such as Nedstack and Nuvera are in the process of ramping up their manufacturing capability.
2. Bring costs down to a level beneath which receiving government subsidies is the only feasible route to product adoption. This second point of bringing down the costs is high on the agenda of all large number of companies and they have identified a multi-pronged approach to this task which includes:
   - Increasing manufacturing and switching from batch to process production lines,
   - Decrease the level of component redundancy,
   - Increase stack life and reliability in order to decrease operation and maintenance (O&M) costs, and
   - In some cases, reengineer stack architecture.

Table 6.2 Developer-aimed prices of non-residential stationary fuel cells

Company	Technology	Size (kW)	Year	Price (US$)/kW
UTC	PAFC	200	2010	1500
Siemens	SOFC Hybrid		2010	400*
Rolls Royce	SOFC Hybrid	1000	?	120–300
Hydrogen LLC	PAFC	400	2010	1500 (from 6 to 25 MW)
Nedstack	PEM	100	2010	(waiting for cost confirmation)
FuelCell Energy	MCFC		2010	(waiting for cost confirmation)

* Note that the US Government has a specific programme for SOFC called SECA (Solid Energy Conversion Alliance) with a cost target of US$400/kW by 2010. Siemens is working within SECA towards this target.
MCFC, molten carbonate fuel cell; PAFC, phosphoric acid fuel cell; SOFC, solid oxide fuel cell.

Table 6.2 outlines the company forecasts as to future price points based on the work that they are doing on the technological front and also scale economies of manufacturing.

The main thing to take away from this table is that all companies are anticipating major cost reductions between now and 2010.

6.7 Fuel choice

Fuel choice for this application is based on consumer demand. Anaerobic digester gas, natural gas, biogas, direct hydrogen and even heavier fuels such as kerosene and coal oil can all be used. In terms of impact on efficiency from using different fuel types, what we are seeing is that different fuels are not impacting, more than a percentage point or two, overall efficiency levels.

6.8 Main development companies involved

1. FuelCell Energy is a manufacturer of MCFC power plants. Its core product is its direct fuel cell (DFC) which can range in size from 250 kW in series up to 2.4 MW. Outside of its native USA, it has distributor agreements in Japan and South Korea and supplies its stacks to the well-regarded German firm MTU CFC.

Future development is a DFC/T, which is a hybrid version of the DFC combined with a gas turbine. Using this model, the fuel cell produces around four-fifths of the power, whilst the turbine produces the rest, recovers the waste heat and, in a neat feedback loop, provides the air for the fuel cell. Modelling work done on this system indicates an efficiency approaching 70%, based on the higher heating value (HHV), with 80% efficiency expected to be possible with larger multi-mega watt systems. One point to note about this system is that unlike the other hybrids under development, the DFC/T is being designed to work at ambient pressure. Current sub-MW systems are being tested with the longer-term aim of a 40 MW power plant.

2. Hydrogen LLC is developing a modular PAFC core unit of 400 kW, comprising of 4 × 100 kW stacks. These units can operate in series up to a 6 MW 'Power Island', though the standard product will be a 2 MW unit.

3. GE Energy has a contract from the US Department of Energy (DOE) to develop a SOFC multi-MW gas turbine hybrid system which can operate on coal. Under the 10-year, US$83 million agreement, GE Energy will develop a system with a SOFC/gas turbine hybrid at its heart. The overall integrated gasification fuel cell (IGFC) is aiming at an efficiency of 50%, using gasified coal as its feedstock. The three phases of this very ambitious development programme are:

- develop a design for a 100-MW IGFC power plant,
- demonstrate a proof-of-concept (POC) system and resolve obstacles associated with the development of SOFC, and
- develop and demonstrate an SOFC building block stack for multi-MW system applications.

In terms of current development, GE Energy has recently demonstrated a 5 kW SOFC.

4. MTU CFC Solutions is a German joint venture between RWE Fuel Cells (18.1% of shares) and MTU Friedrichshafen (89.9% of shares) and gets its stacks from FuelCell Energy. (To complete the picture MTU Friedrichshafen is owned by DaimlerChrysler.) The company is now selling its HotModule MCFC CHP unit, which is up and running in a number of locations (some 35 to the end of 2006), primarily in Germany.

5. Nedstack is one of the main European players and is working with PEM technology for both automotive and stationary applications. Since 2006, its

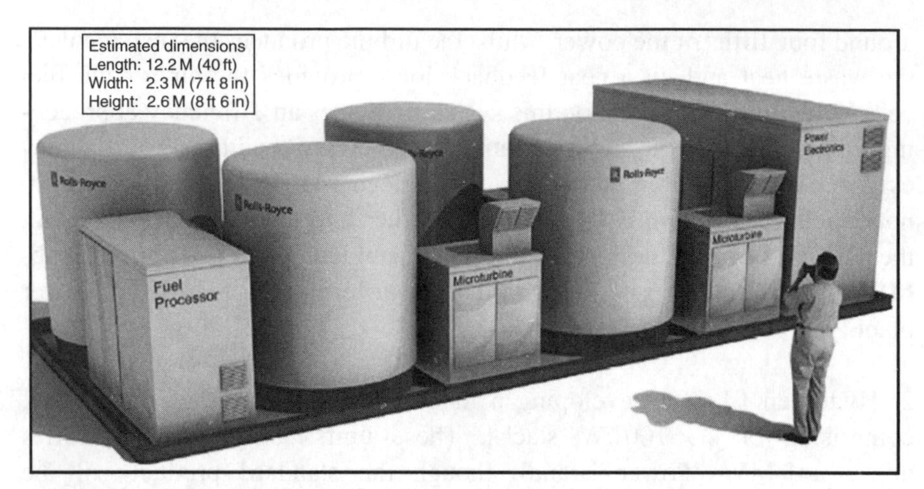

Figure 6.8 Mockup of the Rolls Royce 1 MW SOFC/turbine hybrid system
Source: © Rolls Royce Fuel Cells

work in the area of >10 kW has significantly increased, with the Akzo Nobel project, discussed before, being a real test of its technology.

6. Rolls Royce is working on developing a 1 MW fuel cell/turbine hybrid system, with a demonstration product date of 2008. The system, of which an artist's impression is shown in Figure 6.8, has four 250 kW units linked together.

As well as developing in-house fuel cell capacity, Rolls is also developing its own turbo-generator for the system.

7. Siemens Westinghouse is also working on a high-pressure SOFC/gas turbine design. Though Siemens has been working on the engineering of this system for a number of years, it has yet to come up with an 'optimal' design and states that it is having difficulties with the gas turbines.

8. UTC Fuel Cells is by far and away the largest supplier of PAFC units. Its PureCell 200 (200 kW/925,000 Btu/hr CHP unit) has now been installed in more than 275 sites worldwide and has a reported efficiency of 80% in CHP mode.

One very high-profile project that UTC is involved with is the Verizon call-switching centre in Garden City, New York. This location is home to 1.4 MW (7 × 200 kW units) PAFC power. To date, this is the largest such fuel cell project in the world. The units, which have a low sound profile of only 60 dBa, are being run off natural gas or digester gas and are expected to operate in parallel with the grid, under normal operating conditions, or as primary power under grid failure conditions.

9. Logan Energy. Though Logan is not a fuel cell manufacturer but is a distributor and turn-key project consultant they are worth including here due to the unparalleled level of experience that they hold in this field. They currently hold the title of having installed more units than any other outfit globally and work across the globe on various projects.

Case study – The Freedom Tower

When the Twin Towers of the World Trade Centre came crashing down on 9/11, New York not only mourned the loss of life but also lost a part of its iconic architectural heritage. On 28 June 2006, New York Governor, George E. Pataki, announced a new building that will be built on the site. 'The Freedom Tower' development will in fact be a number of skyscrapers, including The Freedom Tower itself, World Trade Centre Office Towers 2, 3 and 4, as well as the World Trade Centre Memorial and Memorial Museum and will cover an 8-acre site in central Manhattan.

What is innovative about this project, in terms of energy, and why it is included in this book, is that the construction group behind the project have been instructed to use renewable energy and fuel cells to a greater extent than any single project to date. Specifically:

The Freedom Tower and World Trade Centre Office Towers 2, 3 and 4 will be partially fuelled by four 1.2 MW fuel cell systems (4.8 MW in total) to be acquired by the New York Power Authority (NYPA). Also, The Freedom Tower, World Trade Centre Transportation Hub, Memorial and Memorial Museum will be powered by 184 million kW of renewable energy. NYPA will purchase 93 million KW hours of renewable energy credits (RECs) and Silverstein Properties, the developers, will also purchase a further 91 million KW hours of RECs for Towers 2, 3 and 4.

The supplier of the unit has, at the time of writing, yet to be named.

6.9 A list of twenty suggested questions to ask if you are considering using >10 kW fuel cells in a development

General questions

1. Do you have a run down of local interconnection regulations?
2. Have you worked with the local planning authorities from the start of the project?
3. Are you aware of any carbon trading scheme that could be used to help alter the economics?
4. Does the local emergency services need extra training/information in case of emergencies?
5. Are you fully up-to-date on local/government planning regulations?
6. Have you considered the other alternatives?
7. Do you have access to a qualified installer?
8. Do you have access to a qualified maintenance technician?
9. Are you on good terms with the local press?
10. Do you have a lot of patience?

Fuel cell-specific questions

1. Do you want/need high-grade waste heat?
2. Do you know what size unit best suits the developments needs?
3. Do you have a fuel cell supplier for the unit?
4. Do you prefer to use a turn-key project group?
5. Are there any local/national subsidies that may be available to install a fuel cell?
6. Do you have a good supply of fuel?
7. Is there an option for renewables to be linked into the system?
8. Are you wanting to buy or lease the fuel cell?
9. Have you got a thorough breakdown of codes and standards for hydrogen storage (if using direct hydrogen)?
10. Do you feel confident that you have all the information to make the right decision for your project?

6.10 Summary

In general, for adoption of fuel cells, these units offer a fairly straightforward entry proposition. Their current economics are relatively well known, and published, with projections for medium-term targets being realistic with ramping up of production and removal of some of the redundancy levels within the technology. A number of different organisations, from state level in the US, to the central government level, in such countries as Germany, are backing their adoption with removal of a number of barriers to potential entry.

PAFCs may have been first to market, and it did initially appear to observers as if they would create a step-change for the fuel cell industry in general but it did not happen in the time scale envisaged. Since this early period, this sector has gone through a difficult period with companies such as Ballard and Fuji Electric mothballing their plans for units above 10 kW. Now though with the confluence of social, political and technical factors, we are facing a window of opportunity in which these larger units will lead the pack. And where they lead, other applications will follow.

This book has tried to provide a portfolio of information to potential adopters on four main stationary fuel cell applications: residential, UPS, commercial buildings and power plants. Though a number of technological and regulatory barriers still exist, adoption is starting to ramp up, especially for UPS and commercial buildings. Of the four applications, residential is the furthest from full commercial availability but with strong government support it would not be far behind.

Though the basic issue underpinning this entire technological option is the potential to switch to a DG model. So the question is – is the DG model the future energy paradigm? We think so.

7

Government and NGO support programmes

7.1 Introduction

It is a common debate in the fuel cell community as to what governments should, and could, do to support the further development, market introduction and penetration of fuel cell technology.

One side of the debate is that as with many other technologies their fate, in terms of competition with rival technologies, should be left strictly to market forces; the best technology for the market will win through, independent of government intervention. The other side of the debate is that for technologies that could directly contribute to governmental level objectives (e.g. energy security, carbon emission reduction, local air quality and grid stability), some support should be given to at least increase the technology's chances of making a successful diffusion into the open market, e.g. help with crossing the so-called 'chasm of commercialisation'.

There have been times in near history where governments in countries such as the UK and the USA have legislated in favour of the use of certain technologies, such as catalytic converters in new cars. The counter-argument to this technology forcing has always been that if you legislate for one technology today you are legislating out of the market non-complementary technologies that could potentially in the long term have more beneficial effects. In the case of the catalytic converter, one indirect impact of legislating for the use of this technology was to slow down the development of lean-burn engine technology, which at that time could not work with three-way catalysts.

This is a massive simplification of a very long running, detailed and heavy (and fascinating) debate on the role of government and technology adoption/diffusion. Suffice to say that governments have a difficult job and that for fuel cells, interestingly, we see approaches from both ends of the spectrum.

In Japan and Korea, fuel cells have been directly identified and targeted as a technology that the governments see contributing to their long-term objectives. To help with the development not only of the technology but also of their corresponding markets, both countries have put in place very specific support programmes, outlined later in this chapter. The planned impact of these programmes is market introduction of fuel cells in a number of applications within the next decade within these regions. The USA has taken a different approach and is currently investing heavily into technological research. By developing a range of very specific target metrics, it believes that the technology will then be ready to face a competitive market place with minimal governmental support. The European Union (EU) strategy sits somewhere between Japan and the USA. The EU works with a range of instruments to complement policy objectives and for fuel cells whilst the programme does not go as far as its equivalent in Japan, in terms of technological selection, it does provide more support for large-scale demonstration and deployment than does the programme of the US federal government.

Before each of the different policy approaches and targets are outlined in further detail (below), it should be pointed out that governments support fuel cells and hydrogen for different policy objectives. These different foci have helped to create correspondingly different development targets, speed of development and even global relationships.

Whilst the USA, and Australia, is very keen on pursuing the development of the coal to fuel route for use in the fuel cell, countries such as India and China are facing immediate concerns on air quality. Japan and Korea are different again in seeing fuel cells fit into the government objectives on energy security and continued development of their high-tech economies. On this map, the EU stands alone in that one of the drivers behind the interest in fuel cells and hydrogen is the need to reduce greenhouse gases. As well as being a truly fascinating subject, the geopolitics of the interest in fuel cells, and hydrogen, helps create an understanding of why certain policy measures are being driven forward.

The sections that follow are intended to provide an overview of the varying strategic decisions governments around the world have taken towards supporting fuel cells, as well as looking at the motivating factors for doing so.

Asian policies are considered first, then North American and finally European. Whilst the overviews will necessarily be a snap shot of the situation, due to the fact that individual national governments (let alone multi-governmental constructs such as the EU) rarely make homogeneous policies, there are nonetheless strong trends that characterise geographical regions.

7.2 Asia – Japan and South Korea

7.2.1 Japan

The Japanese government, more than any other country globally, has made an explicit decision to support the route to mass market for fuel cells. Not only has it selected fuel cells as a technology that it wants to be adopted and used in its society, but, working together with Japanese industry, it has also produced a highly detailed, step-by-step route map of how it wants to get there.

The current 2006 route map itself is 118 pages long, when translated into English[1], and provides tables of information on research and development (R&D) targets, breakdown on research points and dissemination goals. As well as providing support for the required research, the government has also placed into the market a series of subsidies for the large-scale demonstration of fuel cells.

Japan has without doubt placed a proverbial stake in the ground by pre-selecting the technology that it sees will contribute to the long-term economic and environmental well-being of its country. It has backed this decision heavily with large amounts of funding and other means of support.

As the remit of this text is only stationary fuel cells, the next section outlines only the relevant targets from the route map and provides an overview of the standard-setting Japanese residential fuel cell programme.

2006 New Energy and Industrial Technology Development Organisation (NEDO) roadmap

The route map is split into a number of sections based on application (stationary, vehicular and portable) and technology [primarily direct methanol fuel cells (DMFCs), proton-exchange membrane (PEM) and solid oxide

---

[1] A full translation of the route map can be downloaded from Fuel Cell Today free of charge.

fuel cells (SOFCs)]. Within these sections, the differing components are analysed alongside dissemination targets. Table 7.1 and Table 7.2 outline the key targets for PEM and SOFC technologies for stationary applications.

As can be seen from these tables, not only are the performance metrics for the technology challenging but also, critically for market success, the cost targets are set a level to ensure competitiveness with incumbent technologies in an open market place (Tables 7.3, 7.4).

Japanese residential fuel cell programme

One example, highly relevant to this book, of a specifically targeted programme in Japan is the residential fuel cell programme. Early in 2005, the New Energy Foundation (NEF) announced that it was to make a multi-year large-scale demonstration of small (residential) stationary fuel cells. This demonstration will receive over the 3-year time period substantial subsidies from NEDO for a total of 6400 fuel cells. Current (2006) 1 kW residential fuel cell units cost approximately 10 million yen (US$80,000) each, with the installation costing another 1 million yen per unit (US$8,000). Of the units involved in the programme, the energy companies are required to pay the difference between the actual cost and the subsidy as shown in Table 7.5.

The objectives of this program are to demonstrate residential systems, installed with the following performance vectors:

- approximately 1 kWe output,
- over 30% electric efficiency and over 65% overall efficiency at rated output operation (HHV),
- over 27% electric efficiency and over 54% overall efficiency at 50% load operation over a 2-year lifetime.

Year one of the programme ran with a limited number of companies and a conservative number of units. Year two of the programme, which started late 2006, has seen an extended number of companies involved and a greater number of units. During the programme, the active companies will not only have units in the field being tested but they will also undertake pre-identified and targeted research. For example, Osaka Gas and Sanyo Electric are undertaking work on the verification of test results for natural gas fuelled units, and Showa Shell Sekiyu, Kyushu Oil and Taiyo Oil are working on LPG reformers.

Table 7.1 Japanese stationary proton-exchange membrane (PEM) performance targets

	2006	**2007 (early introduction)**	**2010 (improved model)**	**2015 (dissemination)**	**2020–2030 (full dissemination)**
Power generation efficiency	~32% HHV ~35% LHV	~32% HHV ~35% LHV	~32% HHV ~35% LHV	~34% HHV ~37% LHV	~36% HHV ~40% LHV
Stack durability	~10,000 hours	40,000 hours	40,000 hours	40–90,000 hours	90,000 hours
Operating temperature	~70°C	~70°C	~70°C	~70–90°C	~90°C
System cost	Several million ¥/kW	1,200,00 ¥/kW ~10,000 US$/kW (assuming production of 10,000 items/year)	700,000 ¥/kW ~6000 US$/kW (assuming production of 20,000 items/year/company)	500,000 ¥/kW ~4000 US$/kW (assuming production of 100,000 items/year/company)	<400,000 ¥/kW ~3500 US$/kW (assuming production of >500,000 items/year)

Table 7.2 Japanese stationary solid oxide fuel cell (SOFC) performance targets for small CHP units (<10 kW)

	2006–2010 (system verification)	**2015 (early introduction)**	**2020–2030 (dissemination)**
Power generation efficiency	∼40% HHV ∼44% LHV	∼40% HHV ∼44% LHV	>40% HHV >44% LHV
Stack durability	∼10,000 hours	40,000 hours	90,000 hours
System cost	10,000,000 ¥/kW ∼85,000 US$/kW	1,000,000 ¥/kW ∼8500 US$/kW	300,000–400,000 ¥/kW ∼2500–3500 US$/kW

On the financial side, the companies are split between those who will lease their units for 'free' to the households which will use them and those who will charge a form of rental fee. Osaka Gas is an example of a company not charging, whilst both Tokyo Gas and Nippon Oil charge.

This one programme without doubt will have larger ramifications for the entire residential fuel cell sector. If the targets are met and by 2008 we have a 1 kW residential unit costing under US$8000, then we would be a large step forward towards making this application a commercial reality.

As an aside to this, it would be easy to look at Japan's targets and aims for fuel cells and think that they were nothing more than wishful thinking. But, if you look at the development and uptake of solar power in Japan, you might start to rethink your position.

In 1994, Japan started another residential incentive program, this time to stimulate the uptake of solar roof panels. There was once again a clear target, this time about 70,000 solar homes by 2002. Since the start of this programme, Japan has annually invested over US$115 million in photovoltaics, seeing, in return, a 35-fold increase in installed capacity. Since the start of the programme, the average total system cost has fallen by about 75% in real terms and, some analysts believe, the country is now approaching the point at which government rebates will no longer be needed and can therefore be withdrawn. But, and this is key, the cost competitiveness is not due solely to the decrease of the costs of the solar panels, but by the huge drop in prices in all the other pieces of equipment that go with it, most of which are Japan specific. This know-how and technology therefore would be difficult to export aboard where different formats, codes and standards predominate. For residential fuel cells, therefore, it is not fanciful to contemplate a situation where they are price competitive in Japan yet still expensive elsewhere.

Table 7.3 Japanese stationary solid oxide fuel cell (SOFC) performance targets for medium-capacity CHP units (10–several hundred kW)

	2006–2009 (system verification)	**2009–2013 (system verification)**	**2013–~2020 (early introduction)**	**2020–2030 (dissemination)**
Power generation efficiency	~40% HHV ~44% LHV	~40% HHV ~44% LHV	~42% HHV ~47% LHV	>45% HHV >47% LHV
Stack durability	~10,000 hours	10,000–20,000 hours	40,000 hours	90,000 hours
System cost	10,000,000 ¥/kW ~85,000 US$/kW	Several million ¥/kW	1,000,000 ¥/kW ~8500 US$/kW	<200,000 ¥/kW ~1700 US$/kW

Table 7.4 Japanese stationary solid oxide fuel cell (SOFC) performance targets for distributed generation/industrial use (>100 kW)

	2006–2009 (system verification)	**2009–2013 (system verification)**	**2013–2025 (improved model)**	**2025–2030 (full dissemination)**
Power generation efficiency	~50% HHV ~56% LHV	~55% HHV ~61% LHV	~60% HHV ~67% LHV	>60% HHV >67% LHV
Stack durability	~10,000 hours	10,000–20,000 hours	40,000 hours	90,000 hours
System cost	10,000,000 ¥/kW ~85,000 US$/kW	1,000,000 ¥/kW ~8500 US$/kW	Several hundred thousand ¥/kW	<100,000 ¥/kW ~900 US$/kW

Table 7.5 Japanese small stationary subsidies

	2005	2006	2007
Number of units	400	1,000	5,000
Maximu, subsidy (mYen per unit)	6	3	2
Cost target (mYen per unit)	8–10		<1

NB 1 mYen (million Yen) is approximately US$8000.

7.2.2 South Korea

In 1999, after 40 years of sustained high-tech industry growth, which saw South Korea grow from having an economy on par with a number of developing countries to having the world's eleventh largest economy, the Korean government decided that the only sustainable way forward was to transform Korea into a 'science- and technology-based society'.

To enable this transition, the government launched two initiatives – one longer term and one shorter – though feeding directly into the longer term programme. The longer term programme is called *Long-Term Vision for Science and Technology Development Toward 2025*. It has as its core a list of some 40 strategic tasks that will (it is hoped) aid Korea in its transition to the science- and technology-based society by 2025. The second programme was also launched in 1999 and lists technologies to be funded and developed that have the potential, in the medium term, to bear fruit and help support the 2025 Vision. This second programme is called the *21st Century Frontier Science Programme*. Each of the technologies and projects being developed in the Frontier 21 programme are for a 10-year period and includes hydrogen energy. Fuel cell research, on the other hand, comes under a subsection of a *10-Year Plan of Energy Technology R&D*, which includes the '*National Fuel Cell Technology Plan*'.

The two main government agencies involved in Korea on both hydrogen and fuel cells are the Ministry of Science and Technology (MOST) and the Ministry of Commerce, Industry and Energy (MOCIE). Between them, they formed the National RD&D Organisation for Hydrogen and Fuel Cells to promote and coordinate the R&D efforts in this area. Also, under the auspices of the 21st Century Frontier Science Programme, in 2003 MOST launched the Hydrogen Energy R&D Centre.

Historically, fuel cell research in Korea has focused on molten carbonate fuel cells (MCFCs) and phosphoric acid fuel cells (PAFCs) for large stationary

Table 7.6 Historical stationary fuel cell development in South Korea

FC type	Date	R&D	Organisations involved
PAFCs	1985–1991	6 kW testing	KIER, KEPRI and KEPCO
	1988–1992	Development of 2 kW system	KIER, MOST
	1993–2000	Development of a 50 kW system	KIER, LG Caltex, MOCIE
PEMFCs	1994–1995	Development of 1 kW system	KOGAS
	1996–2001	Development of 5 kW system	KIER, MOCIE
	2002–	Development of 3 kW system	KIER, CETI, MOCIE
MCFCs	1993–1996	Operating and experience of a 2 kW system	KEPRI, KIST, MOCIE
	1997–2000	Operating and experience of a 25 kW system	KEPRI, KIST, MOCIE
	2001–	Operating and experience of a 100 kW system	KEPRI, KIST, MOCIE
SOFCs	1994–1998	Operating and experience of a 100 kW system	Ssangyong, MOCIE
	2002–	Operating and experience of a 10 kW system	KEPRI, MOCIE
	2003–	Operating and experience of a 1 kW system	KIER, MOCIE

CETI – Clean Energy Technologies, Inc., www.ceti-fuelcell.com.
KEPCO – Korean Electric Power Corporation, www.kepco.co.kr.
KEPRI – Korean Electrical Power Research Institute, www.kepri.re.kr.
KIER – Korean Institute of Energy Research, www.kier.re.kr.
KIST – Korean Institute of Science and Technology, www.kist.re.kr.
KOGAS – Korean Gas Corporation, www.kogas.or.kr.

applications, with Table 7.6 providing a snapshot of the work that has been done so far and by which organisations.

Since the government decided that fuel cells are one of the key technologies for Korea's future growth, the R&D efforts have broadened in scope to include transport and mobile applications, and have also refocused most of the effort on PEM and SOFC formats.

Interestingly, as can be seen from the forthcoming pages, the R&D targets in South Korea are more focused on the applications, rather than on the technologies making up the product. This does allow for an approach which is highly focused on creating a commercial product whilst simultaneously keeping the technological options open.

Figure 7.1 provides a translation from the Korean Fuel Cell R&D Roadmap, showing the targets, up to 2010, for stationary applications. This is followed by a more detailed overview on each area.

Korea has drawn together a number of applications, such as combined heat and power (CHP) units for the home, uninterruptible power supply (UPS)/backup power generation and large-scale power plant fuel cell units.

The targets for this broad application area are:

- Power generation efficiency
  : >40%
  : >80% in energy use when using heat
- Durability
  : >40,000 hours for non-consecutive power use
  : >10,000 hours of cumulative operating life cycle, when operating 5 hour/day
- Price
  : <US$ 1500 kW at commercialisation
- Capacity: Private generation 10 kW, or less
  : Medium–large 100 kW–MW (MCFCs, PAFCs and SOFCs)

Although details on the actual research programmes for fuel cells are not as detailed as they are for hydrogen, it is known that companies such as Samsung (PEM and DMFCs) and Hyundai [fuel cell vehicles (FCVs)] are heavily involved, as well as companies such as LG-Chemical [membrane electrode assembly (MEA) and DMFCs], LG-Electronics (PEM and DMFCs), CETI (PEM and small RPG), Fuel Cell Power (MEA) and SK Corporation (Reformer technology). KIER is working on PEM, SOFCs, DMFCs and PAFCs and KIST is looking into MCFCs, PEM, DMFCs and SOFCs.

In terms of demonstration and technology dissemination, Table 7.7 highlights the main medium-term targets for both stationary and transportation fuel cells.

South Korea, in a similar way to Japan, has made a very clear commitment to fuel cells, alongside wider hydrogen-based objectives. Long-term plans have been made with foundations being laid by both industry and government working together.

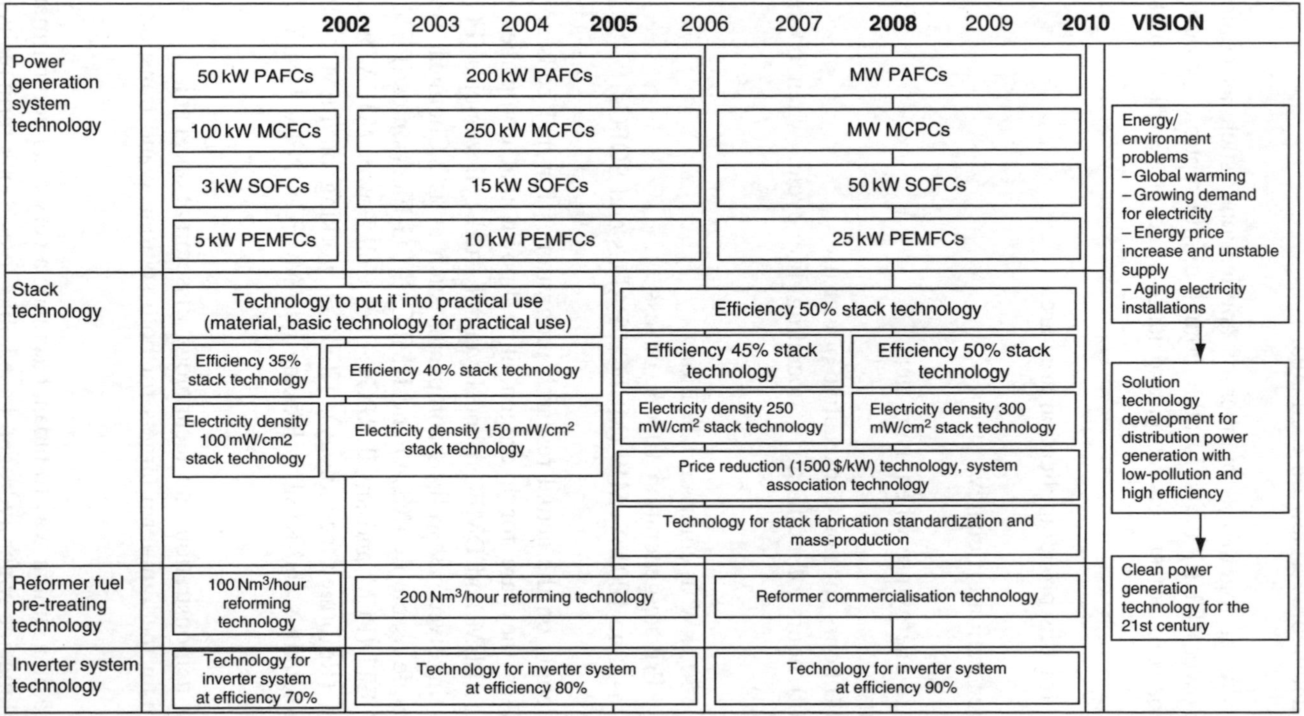

	2002	2003–2005	2006–2010	VISION
Power generation system technology	50 kW PAFCs	200 kW PAFCs	MW PAFCs	Energy/ environment problems – Global warming – Growing demand for electricity – Energy price increase and unstable supply – Aging electricity installations
	100 kW MCFCs	250 kW MCFCs	MW MCPCs	
	3 kW SOFCs	15 kW SOFCs	50 kW SOFCs	
	5 kW PEMFCs	10 kW PEMFCs	25 kW PEMFCs	
Stack technology	Technology to put it into practical use (material, basic technology for practical use)		Efficiency 50% stack technology	↓
	Efficiency 35% stack technology	Efficiency 40% stack technology	Efficiency 45% stack technology → Efficiency 50% stack technology	Solution technology development for distribution power generation with low-pollution and high efficiency
	Electricity density 100 mW/cm2 stack technology	Electricity density 150 mW/cm^2 stack technology	Electricity density 250 mW/cm^2 stack technology → Electricity density 300 mW/cm^2 stack technology	
			Price reduction (1500 $/kW) technology, system association technology	
			Technology for stack fabrication standardization and mass-production	↓
Reformer fuel pre-treating technology	100 Nm^3/hour reforming technology	200 Nm^3/hour reforming technology	Reformer commercialisation technology	Clean power generation technology for the 21st century
Inverter system technology	Technology for inverter system at efficiency 70%	Technology for inverter system at efficiency 80%	Technology for inverter system at efficiency 90%	

Reference: MOCIE, "Fuel Cell Technology Roadmap" "위원회 명단", available from: http://kric.kist.re.kr/pages/Future%20Technology/techroadmap (in Korean).

Figure 7.1 South Korea stationary fuel cell roadmap

Table 7.7 Korean medium-term fuel cell dissemination targets

	Phase 1 (2003–2005) R&D	**Phase 2 (2006–2008) Dissemination**	**Phase 3 (2009–2012) Penetration and market enlargement**
Hydrogen stations	1	10	50
Distributed power		Cumulative 300 units (250–1000 kW)	
Buildings		Cumulative 2,000 units (10–50 kW)	
Residential power		Cumulative 10,000 units (<3 kW)	
Transportation	10 Light-duty FCVs	1,000 light-Duty FCVs 100 buses	10,000 Light-duty FCVs 5000 buses
Portable power	Development of key technologies for commercialisation	Commercialisation of each product	

FCVs, fuel cell vehicles.

7.3 North America

Unlike Japan and South Korea, the USA has a high level of governmental federalisation. Within this, a two-tier system has emerged where R&D activities are carried out mainly at the federal (central) government level, whilst demonstration and dissemination is being done at state level.

In terms of fuel cells and hydrogen R&D, there are two key federal departments, the Department of Energy (DoE) and the Department of Defense (DoD).

The DoE is comprised of four sub-departments. Three of these departments, known as Offices, the Office of Energy Efficiency and Renewable Energy, Office of Fossil Energy, and the Office of Nuclear Energy, Science and Technology, are central to the setting of a number of targets for hydrogen and fuel cells.

Currently, there are three R&D areas with specific aims and objectives: hydrogen production, from a range of current and future sources; FreedomCar; and a programme on fuel cell research. These are grouped together into two specific strands – (1) The Hydrogen, Fuel Cells and Infrastructure Technology Program and (2) FreedomCar and Vehicle Technologies Program.

The Hydrogen, Fuel Cells and Infrastructure Technologies Program provides an umbrella for a number of fairly separate, and often seemingly disparate, research activities. It integrates the research in hydrogen production, storage and delivery with those in transportation and stationary fuel cells. Most of the work being undertaken in this programme has fairly specific aims and targets.

7.3.1 Hydrogen, fuel cells and infrastructure technology program

As mentioned previously, the Hydrogen, Fuel Cells and Infrastructure Technology Program is the umbrella programme which draws together the R&D aims and targets within the relevant DoE departments. Within this programme, R&D aims and targets are set for hydrogen production, distribution and storage, and fuel cell development targets are set for stationary and mobile applications.

The programme has a large number of focused targets, especially for hydrogen production, storage and distribution. As the focus of this text is stationary fuel cells, the main targets of relevance are the two aims to reduce the cost and increase the efficiency for both transportation and stationary fuel cell applications.

Specific targets of interest are:

1. Target:
   - 2010 – $45/kW and 2015 $30/kW for a 40% efficient fuel cell + reformer (using clean hydrocarbon and/or biofuel) for transportation with a startup time of 30 seconds
   - 2010 – $400–750/kW for PEM system for distributed generation, fuelled by natural gas/propane, with 40% electrical efficiency and 40,000 hours durability
2. Aim – by 2015 to ‘develop and demonstrate the critical high-risk technology advancements that will permit the US industry to establish commercial availability of advanced, low-cost, ultra-high-efficiency, fuel flexible, integrated fuel cell/turbine hybrid systems for synfuel and hydrogen-based plants’.

This last aim is very interesting as it stands in sharp contrast to the other targets as being basically immeasurable, potentially leading to difficulties in proving the level of attainment.

From the recently updated Hydrogen Posture Plan, it is clear that in terms of federal support hydrogen production pathways and fuel cell automotive technologies are the two key areas that are being currently funded for R&D. For those interested in a full run-through of all the specific hydrogen targets, the updated 2006 Hydrogen Posture Plan is fully referenced at the end of the book.

As the level of state autonomy in the USA is high, it is important when considering policy developments to also look at state activity. Many of the states have their own renewable portfolio standard (RPS), grid interconnection standards and net metering. Currently, fuel cells, whether fuelled by fossil or renewable energy derived hydrogen often fall into these brackets. Leading state proponents of fuel cell technology include California, Connecticut, New York and Ohio.

7.4 Europe

The EU is a geopolitical grouping of (currently) some 25 member countries. Whilst each country retains sovereign powers over its nation state, a number of policies, increasingly those in the energy field, are made at European level.

7.4.1 EU FP7

For fuel cells, the most relevant are the framework programmes. The framework programmes focus on technologies that could help broad EU energy policy objectives in the medium to long term, with fuel cells being one of the key technologies identified as having a role to play.

The seventh Framework Programme of RD&D was officially launched in late 2006 and runs for 7 years. During this period, some 2.3 billion euros has been ring-fenced for energy research. Within this money, all energy technologies bid for money based on consortia level projects. The bids are drawn together from a pre-selected list of topics that the EU stipulates. FP7 focuses on the medium to long term, with an objective to look at step-change technologies. Within fuel cells, and of interest to the scope of this book, is the funding that could be available for low-carbon buildings.

Fuel cell and hydrogen JTI

New to FP7 is a funding mechanism known as the joint technology initiative (JTI). The JTI is a way of industry and the EU government working together to develop an R&D programme. The programme is designed to be target orientated with a long-term vision. Currently, the JTI for fuel cells is to be confirmed. It is expected that at some point during 2008, the regulations to allow the JTI to kick-off will be passed and the research programme will be launched. Unlike in an open competition where all relevant technologies compete for a sum of money, the JTI will have a ring-fenced pot of funding specifically for hydrogen and fuel cells. This money will be drawn from the total energy budget. At this stage, the total worth of the fuel cell JTI has not yet been announced.

It is anticipated that once the JTI does commence, the work programme will be based on the Hydrogen and Fuel Cell Technology Platform (HFP) Implementation Plan (IP). As this will probably be the basis of the JTI working programme, a brief outline of the IP is given here.

The HFP implementation plan

The main focus of the plan is on four, so-called, innovation and development actions. These are:

1. Hydrogen vehicles and infrastructure: developing vehicle and infrastructure technologies to kick-start commercialisation by 2015 or earlier
   - emphasis on cost reductions, vehicle component (including PEMFCs and hydrogen storage) and infrastructure build-up
2. Sustainable hydrogen supply: supplying 10–20% of hydrogen energy demand with carbon dioxide-free or lean technologies by 2015
   - focus on low temperature electrolysers in the short term and on carbon dioxide-free of lean centralised mass production technologies in the long term; enable decisions and planning for sustainable hydrogen infrastructure build-up
3. Fuel cells for CHP and power generation: having more than 1 GW capacity in operation by 2015
   - address developments on all three – PEM, MCFC and SOFC – technologies in a balanced way to meet both transition and long-term goals
4. Fuel cells for portable and early markets: bringing ‘thousands’ of fuel cell products in the market by 2010
   - foster and facilitate the introduction of marketable products and industrialisation; select market segments underlying Europe’s strengths

Focusing specifically on stationary fuel cell targets, the overarching objective is to bring fuel cell and electrical power generators to commercial readiness by 2010. This is backed up by a quantitative target of 1 GW capacity by 2015. The 1 GW is based on the cumulative units in a number of areas over MCFC-, PEM- and SOFC-based fuel cells. The target breaks down as:

- 1–10 kW units for residential applications – 80,000 units (50% split between PEM and SOFCs)
- 10 kW–1 MW units for industrial applications – 2600 units (primarily MCFCs and SOFCs)
- >1 MW units for industrial applications – 50 units (primarily MCFCs and SOFCs)

Interestingly, the IP does not include any reference to PAFC technologies. This could be because PAFC units are already commercially available and are therefore not seen to need the R&D support of the other technologies.

Within the EU, a number of countries have their own separate fuel cell and hydrogen development programmes. Germany and the UK are two main driver countries here. Activity in Germany is outlined below.

7.4.2 Germany

During 2006, the German Hydrogen and Fuel Cell Technology Innovation Programme was launched. Together with the fifth Federal Research Programme (launched in 2005), which has a key focus on fuel cell and hydrogen technologies, these programmes will provide over 500 million euros in funding over the next 10 years. Whereas the focus of the research programme is research (no surprises here) into cost reduction, increase of lifetime and reliability, the new innovation programme focuses on market preparation and lighthouse projects to bridge the gap between the laboratory prototypes and the market.

Figures 7.2 and 7.3 show route maps relevant to the scope of this book that have been outlined under this new innovation programme.

7.5 Non-governmental organisations

There are, naturally, a number of other countries and NGOs supporting development and route-to-market for stationary fuel cells. Some are working on government-sponsored research, whilst other nations are lining up to be amongst the early adopter nations. Interestingly, the World Bank has a large-scale demonstration programme underway to place stationary fuel cell units in developing nations.

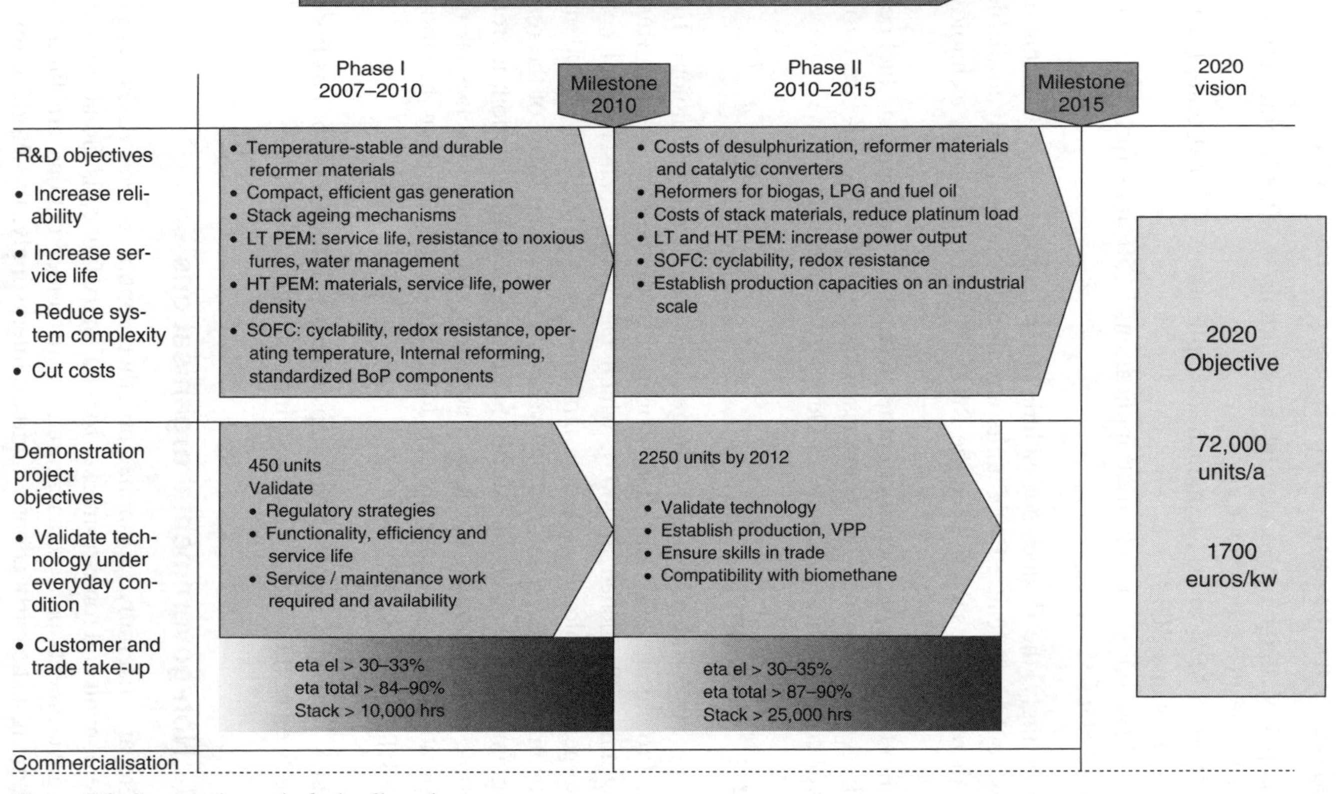

Figure 7.2 German domestic fuel cell roadmap

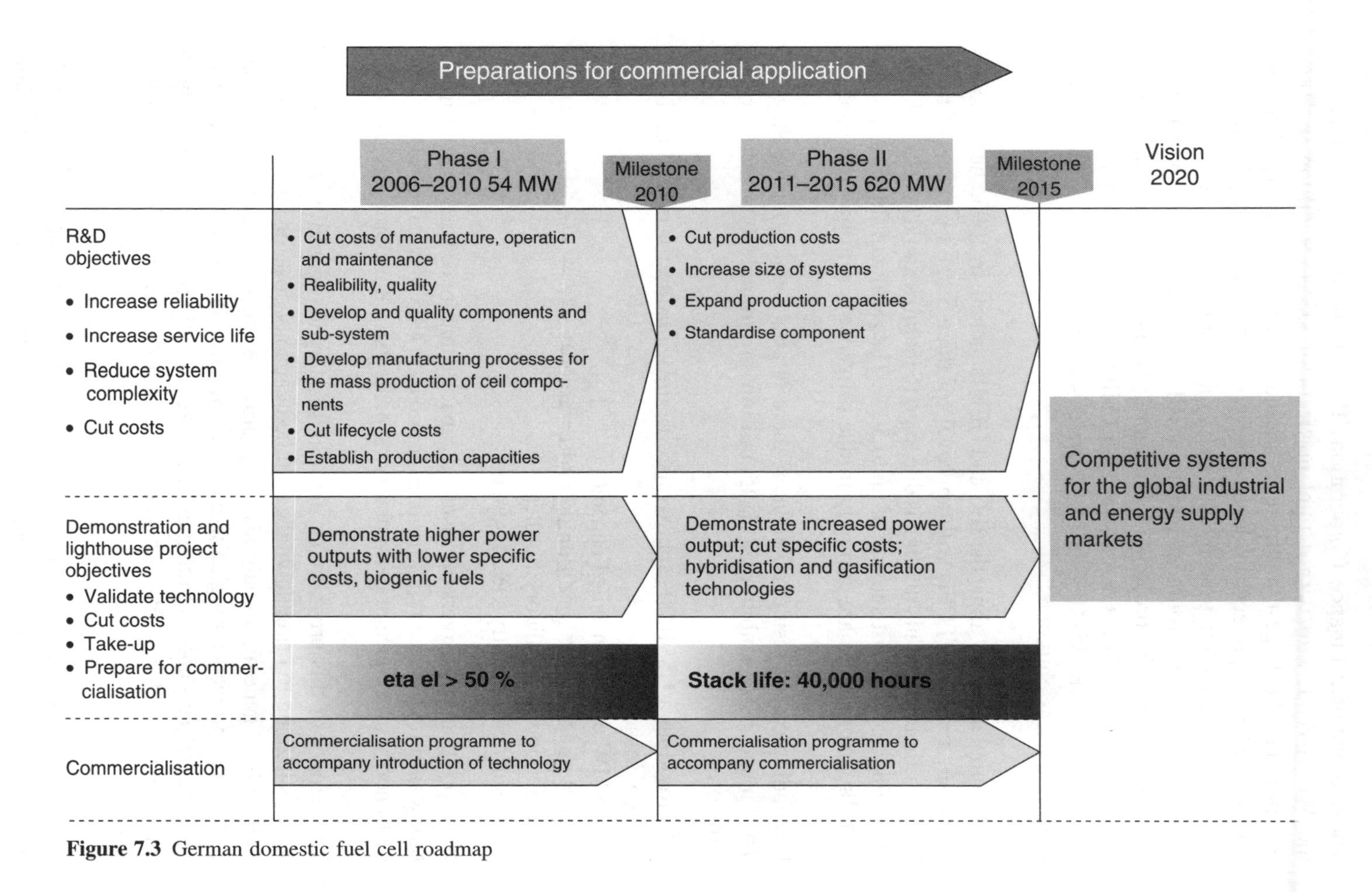

Figure 7.3 German domestic fuel cell roadmap

The International Finance Corporation (IFC) is the private-sector arm of the World Bank and, with a total portfolio of US$17.9 billion in 2004, it represents the largest multilateral source of loan and equity financing for private-sector projects in developing countries. Currently, the IFC has five key priorities for investment, including sustainability. Alongside the IFC, the Global Environment Facility (GEF) provides donor financing for incremental costs for projects in developing countries with global environmental benefits. Of the GEF's total US$3 billion budget, for the fiscal years 2003–2006, some 35–40% is ring-fenced for energy projects that reduce greenhouse gas emissions.

With complementary aims, the IFC and GEF launched, in 2003, the Fuel Cell Financing Initiative (FCFI) with approval for three Stage I projects of up to US$9 million each. This funding is due to go to mainly capital cost buy-down, providing a capital cost subsidy of up to 50%. The results of this first call are that three companies are currently in due diligence for proposed projects.

Depending on the results of Stage 1, there is also a potential for Stage II. This much more ambitious US$45 million programme has the, current, aims of:

1. providing a US$1000 (approximate figure) subsidy for capital costs for new installations
2. 45 MW of new installed capacity
3. providing market conditioning and awareness
4. reduction in regulatory hurdles
5. ensuring in-country O&M support

If the programme delivers the installed capacity that it is currently looking at, this would represent a total, for both rounds, of 49.5 MW of stationary fuel cell energy in a concentrated number of projects in developing countries.

The benefit of this programme, if it hits the targets, would be twofold. As well as helping to bring down the installed capacity cost for large stationary fuel cells in general, it could also help draw developing countries into fuel cell energy projects through, for example, the clean development mechanism (CDM). Projects and programmes like these not only help to raise awareness but provide real benefits and generate a number of beneficial spill-over impacts.

In terms of pure adopter nations, the Indian energy roadmap has made it clear that it will be an adopter nation for fuel cells but in the short term is investing in hydrogen technology for use in an internal combustion engine. Once fuel cells satisfy/meet certain key metrics, in terms of cost and durability, then the government will move to start dissemination activities.

7.6 Summary

A number of countries globally have put their weight behind fuel cell technology, whether as in the dilute, market-based support, case of the UK, by altering the building regulations to strongly recommend the consideration of low- or zero-carbon technology, or as in the other end of scale Japan, which has stated outright that it wants specific numbers of stationary fuel cells installed and operational by a set date. Global momentum behind the technologies is substantial. As well as countries such as the USA, which at the federal level is very heavily focused on R&D, there are a further group of countries that are aligning themselves to be future adopter nations.

In summary, fuel cell technology is one of a basket of technologies that are being developed to help counter the government- and market-based concerns of energy security, urban pollution and climate change.

What we could well see in the coming decades is a proliferation of fuel cells in various buildings, from homes to museums to high-rise office blocks and acting as decentralised power plants. The investment required to realise this vision – in terms of money, political will and ingenuity – is immense. However, the stakes could not be higher. Humankind's natural desire for economic and social progress will surely continue, but without recourse to new, benign energy technologies, we are unlikely to progress far. The promise of fuel cells, though often misrepresented, remains one of our main opportunities for a sustainable future.

Further reading

1. Breakthrough Technologies, "State Activities that Promote Fuel Cells and Hydrogen Infrastructure Development", 2006. Available at: www.fuelceltoday.com
2. NEDO, "2006 Fuel Cell/Hydrogen Technology Development Roadmap", 2006. English translation available at www.fuelcelltoday.com
3. US Department of Energy and US Department of Transport, "Hydrogen Posture Plan: An Integrated Research, Development and Demonstration Plan", 2006. Available at: www.hydrogen.energy.gov

Bibliography

This contains a list of all texts consulted during the production of this book:

Adamson, K.-A., (2005) "Korean Fuel Cell and Hydrogen R&D Targets and Funding", Fuel Cell Today, www.fuelcelltoday.com

Adamson, K.-A., (2006) "Fuel Cell Today Small Stationary Survey 2006", Fuel Cell Today, www.fuelcelltoday.com

Adamson, K.-A., (2006) "Fuel Cell Today Large Stationary Survey 2006", Fuel Cell Today, www.fuelcelltoday.com

Adamson, K.-A., (2006) "Fuel Cell Today Niche Transport Survey 2006", Vol 1 and Vol 2, Fuel Cell Today, www.fuelcelltoday.com

APC, "Alternative Power Generation Technologies for Data Centres and Network Rooms", APC White Paper #64, 2003

Banwell, P., (2006) "Emerging Technologies: Residential Cogeneration", presented at ACEEE MT Symposium, Long Beach, California, Oct 26–27

Binder, M., (2006) "Phosphoric Acid Fuel Cells" presented at California Energy Commission Fuel Cell Workshop, Sacramento, California, May 31

BP, (2006) "BP Statistical Review of World Energy 2005", available from: http://www.bp.com/liveassets/bp_internet/globalbp/globalbp_uk_english/reports_and_publications/statistical_energy_review_2006/STAGING/local_assets/downloads/pdf/statistical_review_of_world_energy_full_ report_2006.pdf

Brdar, D., Bentley, C., Farooque, M., Oei, P., Rauseo, T., (2006) "Stationary Fuel Cell Power Plant Status", FuelCell Energy White Paper

Breakthrough Technologies, (2006) "State Activities that Promote Fuel Cells and Hydrogen Infrastructure Development", 2006. Searchable database at: www.fuelcells.org/info/statedatabase.html, and report available at: www.fuelceltoday.com

Cockroft, J., Kelly, N., (2006), "A Comparative Assessment of Future Heat and Power Sources for the UK Residential Sector", Energy Conversion and Management, Vol. 47 pp. 2349–2360

Crawley, G., (2006) "Fuel Cell Today Light Duty Vehicle Survey 2006", Fuel Cell Today, www.fuelcelltoday.com

Crawley, G., (2006) "Fuel Cell Today Bus Survey 2006", Fuel Cell Today, www.fuelcelltoday.com

Crawley, G., (2006) “AFC Technology, Status and Development”, Fuel Cell Today, www.fuelcelltoday.com

Crawley, G., (2007) “DMFC Technology, Status and Development”, Fuel Cell Today, www.fuelcelltoday.com

Crawley, G., (2007) “MCMFC Technology, Status and Development”, Fuel Cell Today, www.fuelcelltoday.com

Crawley, G., (2007) “SOFC Technology, Status and Development”, Fuel Cell Today, www.fuelcelltoday.com

Crawley, G., (2007) “PAFC Technology, Status and Development”, Fuel Cell Today, www.fuelcelltoday.com

Crawley, G., (2006) “PEM Technology, Status and Development”, Fuel Cell Today, www.fuelcelltoday.com

Crozier-Coles, T., Jones, G., (2002) “The potential for micro CHP in the UK”, report for the Energy Savings Trust

David, P., “The PureCell Fuel Cell Power Plant: A Superior Co-Generation Solution from UTC Power”, presented at 2006 Fuel Cell Seminar, Honolulu, Hawaii, 13–17 November

Ernst, W., Rodriguez, W, Intwala, K.F., (2006) “Back-up/Peak Shaving Fuel Cells” presentation to DoE FY2006 Annual Progress Report

Federal Energy Management Programme, (2005) “Fuel Cells in Backup Power Applications”, US Department of Energy, DOE/EE-0310, available from: http://www.osti.gov/bridge/

Föger, K., (2006) “Clean Power for Your Home: Technical Challenges and Solutions for a Market Ready Product”, presented at 2006 Fuel Cell Seminar, Honolulu, Hawaii, 13–17 November

GoldenGate Software, (2005) “Developing Contingency Plans for the Recovery of Critical Business Functions”, available from: http://whitepapers. techrepublic.com/whitepaper.aspx?docid=83712

Hoogers, G., (2003), “Fuel Cell Technology Handbook”, CRC Press, Florida

Hydrogenics – product brochures, available from www.hydrogenics.com

Idatech – product brochures, available from www.idatech.com

International Energy Agency, (2006) “World Energy Outlook 2006”, OECD/IEA, Paris

IPCC, (2006) “Climate Change 2007: The Physical Science Base. Summary for Policy Makers”, available from: http://www.ipcc.ch/SPM2feb07.pdf

Koehler, T., (2005) “Fuel Cells in Backup Power Applications”, US Department of Energy, Office of Energy Efficiency and Renewable Energy, available from: www.eere.energy.gov/femp/

Lipman, T.E., Edwards, J.L., Kammen. D.M., (2004), “Fuel Cell System Economics: Comparing the Costs of Generating Power with Stationary and Motor Vehicle PEM Fuel Cell Systems”, Energy Policy, Vol. 32, pp. 101–125

Malinowski. P., Menzen, G., (2006) "Fuel Cells and Hydrogen in the German Energy Research Programme", presented at 2006 Fuel Cell Seminar, Honolulu, Hawaii, 13–17 November

Morgan, R.E., Devriendt J.M., Flint, B., (2006) "Micro CHP – A Mass Market Opportunity?", Ceres Power White Paper

MTU CFC – product brochures, available from www.mtu-friedrichshafen.com/cfc/

NEDO, (2006) "2006 Fuel Cell/Hydrogen Technology Development Roadmap", 2006. English translation available at www.fuelcelltoday.com

Nishikawa, S., (2006), "Current Status of the Large Scale Stationary Fuel Cell Demonstration Project in Japan", presentation to IPHE

Pehnt, M., Ramesohl, S., (2003) "Fuel Cells for Distributed Power: Benefits, Barriers and Perspectives" WWF and Fuel Cell Europe

Perry, M., Strayer, E., (2006) "Fuel Cell Based Back-up Power for Telecommunication Applications: Developing a Reliable and Cost-Effective Solution", published in INTELEC 2006 proceedings

Plug Power – product brochures, available from www.plugpower.com

Rittal – product brochures, available from www.rittal.com

Relion – product brochures, available from www.relion.com

Ruyssevelt, P., Burton, S., (2005) "Low or Zero Carbon Energy Sources – Report 4 Final Report", Office of the Deputy Prime Minister, London

Samuelsen, S., (2004) "Draft Fuel Cells Economic Report" California Air Resources Board

Singhal, S.C.S., Kendall, K., (2003) "High Temperature Solid Oxide Fuel Cells: Fundamentals, Design and Applications", Elsevier, Oxford

Slowe, J., (2006) "Micro-CHP to Increase Energy Efficiency: Emerging Technologies" white paper Delta Energy and Environment

Solar Energy Laboratory, (2001), "Assessment of Solid Oxide Fuel Cells in Building Applications: Phase 1: Modeling and Preliminary Analyses", Energy Centre of Wisconsin

Stern, G., (2006) "The Coming Fuel Cell Revolution", EnergyBiz Magazine, available from: www.energycentral.com

Stern, N., (2006) "The Economics of Climate Change", 2006, available from: http://www.hm-treasury.gov.uk

Teichroeb, D., (2005) "Enabling Clean Infrastructure Opportunities: Stationary Fuel Cells Part of Ontario's Diversified Electricity Portfolio" White Paper Fuel Cells Canada

Tiax, (2002), "Grid-Independent Residential Fuel-Cell Conceptual Design and Cost Estimate", Department of Energy, available from: http://www.osti.gov/bridge/

Torrero, E., McClelland, R., (2002) "Residential Fuel Cell Demo Handbook: National Rural Electric Cooperative Association Cooperative Research Network", National Renewable Energy Laboratory, available from: http://www.osti.gov/bridge/

Torrero, E., McClelland, R., (2004) "Evaluation of the Field Performance of Residential Fuel Cells", National Renewable Energy Laboratory, available from: http://www.osti.gov/bridge/

US Army Core of Engineers, (2004) Engineer Research and Development Centre, "Proton Exchange Membrane Fuel Cell Demonstration of Domestically Produced PEM Fuel Cells in Military Facilities", W9132T-04-C-0017

US Department of Energy and US Department of Transport, (2006) "Hydrogen Posture Plan: An Integrated Research, Development and Demonstration Plan". Available at: www.hydrogen.energy.gov

World Commission on Environment and Development, (1987) "Our Common Future", Oxford Paperbacks, Oxford

Author's Note: Please accept my apologies for any missing texts, this is purely accidental and not intended as a mark of disrespect to the authors.

Index

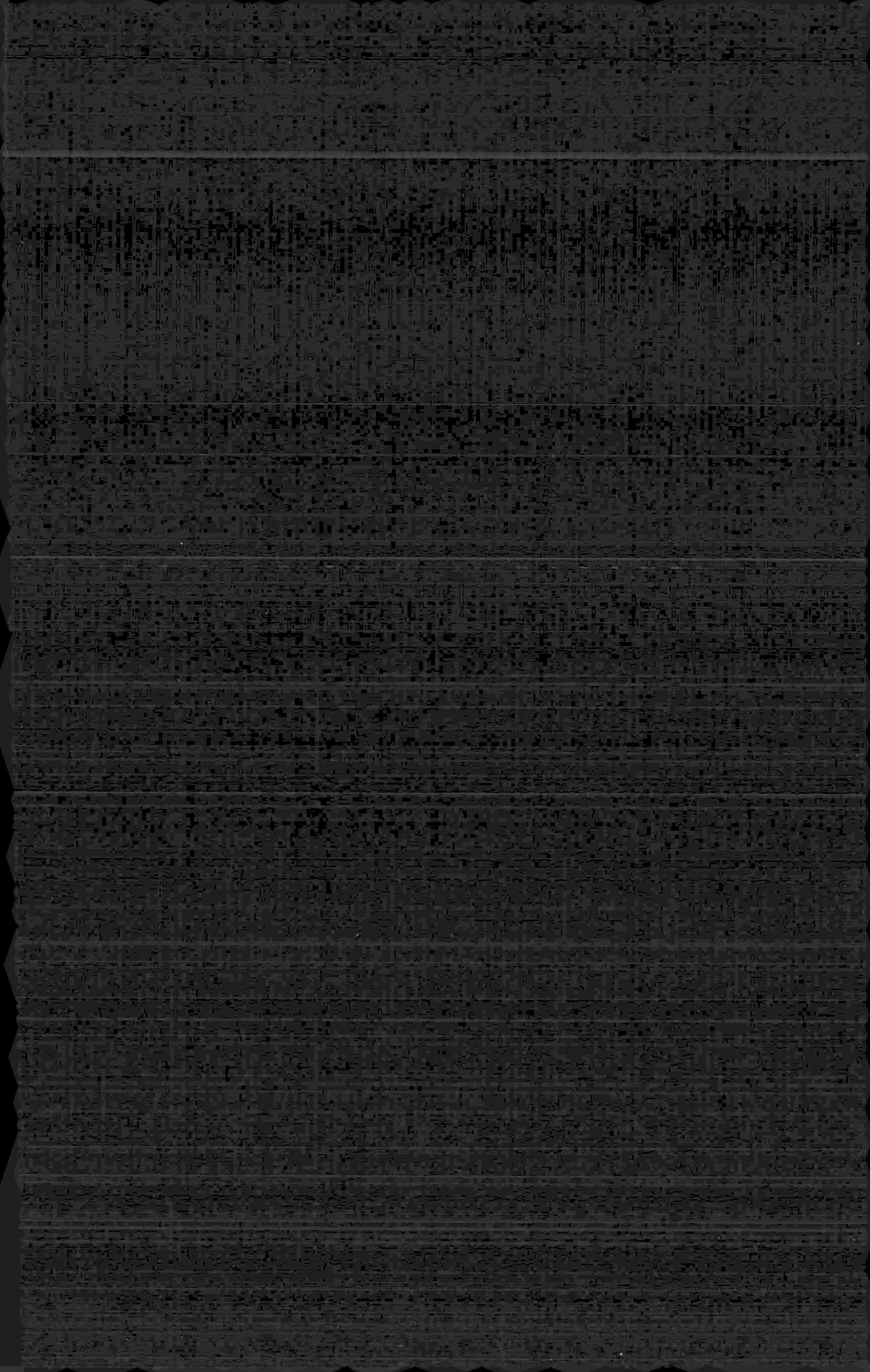